Understanding SIP Servlets 1.1

For a complete listing of the *Artech House Telecommunications Series*,
turn to the back of the book.

Understanding SIP Servlets 1.1

Chris Boulton
Kristoffer Gronowski

ARTECH
HOUSE

BOSTON | LONDON
artechhouse.com

Library of Congress Cataloging-in-Publication Data
A catalog record for this book is available from the U.S. Library of Congress.

British Library Cataloguing in Publication Data
A catalogue record for this book is available from the British Library.

Cover design by Igor Valdman

ISBN 13: 978-1-59693-428-3

© **2009 ARTECH HOUSE**
685 Canton Street
Norwood, MA 02062

10 9 8 7 6 5 4 3 2 1

Contents

Foreword

When SIP was first specified, an important fraction of the IETF community immediately got excited about the possibilities the new protocol promised to open. SIP was based on the protocols behind two of the most successful Internet applications so far: the Web and e-mail. SIP's encoding, transaction, and routing models were based on Hypertext Transfer Protocol (HTTP) and Simple Mail Transfer Protocol (SMTP).

The main reason of all the excitement around SIP was that SIP was designed to facilitate the implementation and deployment of new innovative services on top of it. The fact that SIP was based on HTTP made SIP an excellent service enabler. HTTP was a well protocol on top of which many successful services had been built. Every computer science student was familiar with the technologies used to build services on top of HTTP servers. As a consequence, the number of developers who could develop services on top of SIP using the same technologies was incredibly high.

Soon after SIP was specified, work to adapt technologies to create services on top of HTTP, such as CGI and HTTP Servlets, to a SIP environment started. That is how SIP Servlets were born. Since then, many SIP services have been built using Java technologies in general and SIP Servlets in particular. SIP Servlets provide an abstraction layer that hides some of the complexities of SIP but still allow implementers to directly use SIP primitives in their services. This level of abstraction has proven to be a good match for a wide variety of SIP services.

Chris Boulton and Kristoffer Gronowski do a great job of providing implementers with background information on SIP Servlets, details on how they work, and practical examples that can be run in an open source framework. Both Chris and Kristoffer are well-known and respected members of the SIP community. Their contributions specifying protocol extensions and leading open source development efforts are very much appreciated by the SIP standardization and implementers communities.

Gonzalo Camarillo
Coauthor of SIP (RFC 3261)
Cochair of the IETF SIPPING WG
April 2009

Foreword

SIP Servlets have become a key technology for application development in next-generation networks and are regarded as the primary programming model for SIP-based applications. The concept of bringing familiar, Java-based programming models to SIP, and abstracting away the underlying protocol complexities, opened up the network and significantly shortened the learning curve for developers wanting to build telecom applications. The SIP Servlet 1.0 specification published in 2003 represented an inflection point; the industry was looking for a technology that would enable rapid service creation on the emerging SIP and IMS networks, and SIP Servlets was the first to deliver this.

Much of the initial interest around SIP Servlets came from its relationship with Java and HTTP Servlets specifications. This close bond brought a familiarity that encouraged the adoption of SIP Servlets. However, this was also the center of much debate about the differences between SIP and HTTP, in particular the fact that, although SIP and HTTP are syntactically related, HTTP is fundamentally synchronous while SIP is very much asynchronous. These differences prompted discussion about the suitability of a Servlet component model for SIP development. Ultimately, the developer did need to be aware of the differences; they didn't prove a barrier to the utility of SIP Servlets. A more prevailing issue was the relation of SIP Servlets to other application components. The linkage between HTTP and SIP Servlets promoted the idea of combining SIP Servlets with Web-based applications, but in reality, this was suitable only for simple applications. More important, because of the demand for more complex applications and emerging technologies, there were requirements to allow SIP Servlets to interact with additional constructs such as EJBs, JCA components, JMS queues,

Spring and OSGi components, and so on, as well as to allow SIP Servlet-based features to be exposed as Web services for composition in SOA-based applications.

Addressing such issues, reflecting the changing technology landscape, and incorporating the raft of suggested specifications improvements were key requirements fed into the JSR 289—SIP Servlets 1.1. It was a great pleasure working with Chris Boulton and Kristoffer Gronowski and the rest of the JSR 289 expert group to update the specification. In this book, Chris and Kristoffer do an excellent job of guiding developers through the specification, providing genuine code examples and touching on appropriately related technologies.

James Steadman
Senior Director of Technical Product Strategy at Oracle
Member of the SIP Servlet 1.1 Expert Group
April 2009

Preface

The way we communicate with each other is changing and will continue to evolve in the coming years. The advent of improving core networks has led to various enabling technologies that are helping to maximize the advancements being made. Handheld and fixed-line devices are now considered to be truly multimedia and will continue to converge. The Session Initiation Protocol (SIP) is a catalyst that has propelled IP-based communication over the past 10 years with an increasing number of live deployments and adoption by major technologies such as the IP Multimedia Subsystem (IMS). In 2003, SIP Servlet 1.0 was published to leverage SIP and provide an appropriate application model to rapidly create and deploy new applications and services. It was highly successful as a concept, with many major SIP vendors and adopters using SIP Servlet technology. As with any new technology, implementation experience and real-world application resulted in a number of areas that could be improved in SIP Servlet 1.0. In conjunction with this, the core SIP protocol also evolved, which required SIP Servlet technology to align with it. SIP Servlet 1.1 is the result of such industry collaboration and was officially published in August 2008.

After many years of working with SIP Servlet technology, we authors felt that a supplementary text would be a useful companion for those wanting to learn about the technology and its role. We feel that good SIP Servlet information combined with real-world examples and code snippets will provide readers with a relevant level of knowledge.

The book is split into two main parts. The first part is entitled "Introduction to SIP Servlet Technology" and includes chapters that cover: SIP Servlet Containers, SIP Servlet Applications, Application Routers, and the next directions for the technology.

The second part of the book is entitled "Developer and Deployment Environments" and includes chapters that cover: SailFin technology overviews, SIP Servlet programming, and relationship and role of the technology within IMS.

We hope this book will provide a valuable resource for those in academia and industry who require an in-depth, clear, and concise introduction to SIP Servlet technology. The real-world examples provide helpful aids to those looking to take advantage of SIP Servlets.

Part I
Introduction to SIP Servlet Technology

1

Introduction to SIP Servlets

The telecommunications industry has witnessed dramatic change in recent times with the emergence of Internet Protocol (IP) telephony as a replacement for traditional circuit-switched networks. The evolution toward IP telephony is still in a transitional stage; enabling technologies are being developed and implemented as the telecommunications (telecoms) industry moves toward true multimedia communications. Consumer expectations are constantly increasing as what are becoming multimedia sessions provide not only exchange of traditional voice but also instant messaging, video, gaming sessions, and an unlimited number of new technologies. Fulfilling such a future vision is not an easy task and requires supporting protocols and technologies to provide relevant infrastructure that is interoperable and scalable. The Session Initiation Protocol (SIP) (defined by the Internet Engineering Task Force in RFC 3261[1]) has emerged as the primary protocol solution for IP multimedia communication and has already seen widespread implementation and deployment. It has also been adopted by the Third Generation Partnership Project (3GPP) as the core signaling protocol for its IP Multimedia Subsystem (IMS). SIP Servlet technology is a powerful tool leveraging the multimedia communications that SIP establishes, providing abstracted access to core signaling operations for rapid development of interoperable next-generation services. SIP and SIP Servlets are key enabling technologies that will thrust IP communications into the next phase of evolution.

1.1 Session Initiation Protocol

The Internet has seen an explosion of growth in recent years, which has resulted in worldwide adoption and expansion of related protocols and technologies. The

3

Session Initiation Protocol is such a protocol and was developed in the late 1990s by the Internet Engineering Task Force as RFC 2543 [2]. In the early twenty-first century, the core SIP specification was revised to iron out a number of problems that emerged as a result of early adopters' implementations. RFC 3261 [1] has been the core SIP protocol since and has evolved to become the foundation of fixed, wireless, and Internet communication services. Early adoption of SIP has focused on simple toll bypass, where legacy Public Switched Telephone Network (PSTN) has piggybacked Internet connections for cost-saving purposes, usually across geographical boundaries. It is now clear that SIP is evolving toward its true purpose of creating, managing, and terminating multimedia connections over IP networks.

While RFC 3261 provides the core protocol semantics, the specification was split to provide a suite of extensions for important functionality areas:

RFC 3262 [3]—"Reliability of Provisional Responses in the Session Initiation Protocol." This extension provides the ability to reliably send (since the default for the SIP protocol is unreliable messaging) provisional SIP response messages before a call is connected. This mechanism is especially useful for conveying early call information and, potentially, media such as ringtones or network announcements. It is also used extensively by 3GPP in their IMS architecture for bearer resource reservation in the early stages of call setup.

RFC 3263 [4]—"Session Initiation Protocol: Locating SIP Servers." This extension provides the Domain Name Service (DNS) procedures for SIP when resolving a SIP Uniform Resource Identifier (URI) into physical-network-routing properties such as IP address, transport protocol (e.g., TCP/UDP/SCTP), and port number. This applies to both SIP requests and responses and also provides an inherent failover mechanism in the protocol when multiple entries are configured in DNS for a URI.

RFC 3264 [5]—"An Offer/Answer Model with the Session Description Protocol (SDP)." SDP [6] is used to explicitly describe a multimedia session and is used as a payload in SIP session establishment messages. RFC 3264 provides detailed procedures when using SDP with SIP to create a compatible multimedia session.

RFC 3265 [7]—"Session Initiation Protocol—Specific Event Notification." Subscription and publication of SIP-based events was recognized as an important part of the general SIP infrastructure. This extension provides an extensible framework that uses SIP for requesting and receiving events. Specific usages of the core framework are defined in extension documents that use the core template provided (e.g., presence-based event subscription and notification is defined in RFC 3856 [8]).

RFC 3266 [9]—"Support for Internet Protocol Version 6 (IPv6) in the Session Description Protocol (SDP)." SIP has support for IPv6 inherently built into the protocol, and so as part of the core SIP offering it was recognized that SDP required appropriate extensions to align. This standard specification fulfills that requirement.

The remainder of this section provides an extremely brief introduction to the core SIP specification as defined in RFC 3261. The intention is to provide limited context for this book; those readers who require more in-depth knowledge on the SIP protocol should take a look at the previously mentioned specifications as well as a dedicated book such as *SIP: Understanding the Session Initiation Protocol* [10].

SIP is a revolutionary cleartext application layer protocol that is intended to create, manage, and terminate complex multimedia sessions between entities with potentially different capability sets. The core protocol can be seen as a derivative of the much used Hypertext Transfer Protocol (HTTP) [11], and while they have much in common, there are also significant differences. HTTP is very much based on a client–server relationship in which the server generally issues the final response to a request. SIP, on the other hand, is a routing-based protocol that allows requests to be proxied and redirected for any number of nodes. HTTP also has an extremely rigid client–server model in which an HTTP client issues a request to an HTTP server for a response. SIP has no such constraint, allowing SIP applications and application servers the ability to originate requests.

SIP is media agnostic, so it is not holistically related to internet telephony but is designed for any media interaction, such as multiparty multimedia conferences, instant messaging, video, and whiteboard sharing. As described by RFC 3261 [1], SIP:

> supports five facets of establishing and terminating multimedia communications:
> - *User location:* determination of the end system to be used for communication;
> - *User availability:* determination of the willingness of the called party to engage in communications;
> - *User capabilities:* determination of the media and media parameters to be used;
> - *Session setup:* "ringing," establishment of session parameters at both called and calling party;
> - *Session management:* including transfer and termination of sessions, modifying session parameters, and invoking services.

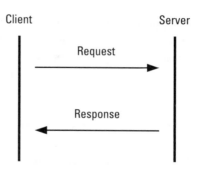

Figure 1.1   Request response interaction.

SIP is a request/response protocol that at the most basic level involves a client issuing a SIP request that will generate a SIP response (as illustrated in Figure 1.1) by a server node.

A slight variation to this theme occurs with the SIP INVITE primitive, which will be discussed later in this section. Due to the end-to-end nature of SIP, it is quite plausible that intervening proxies will not want to remain in the signaling path. Unlike HTTP, SIP can run over both reliable and unreliable protocols [both User Datagram Protocol (UDP) and Transmission Control Protocol (TCP) must be supported by implementations]. UDP is actually defined as the default signaling protocol for SIP, so a three-way handshake is required to ensure multimedia session establishment (as illustrated by the "confirm" message in Figure 1.2).

SIP provides a number of core primitives that are used as request commands for various functions in the establishment and management of multimedia sessions. They are the following (it should be noted that only RFC 3261 primitives are described here, and that subsequent extension primitives that have been created are not covered):

INVITE: The most important primitive defined in the core SIP protocol. The INVITE method is used to initiate and update a multimedia session with another client using the Session Description Protocol.

ACK: The ACK primitive is used in conjunction with the INVITE primitive. It forms the final part of the three-way handshake that is involved in establishing a multimedia session.

CANCEL: After issuing an INVITE request, the transaction will take a varied amount of time to complete. It could be instantaneous or take a longer period of time, maybe in the order of minutes (e.g., as when a telephony call is ringing at one end, but no one is answering the call). The CANCEL request primitive provides the originator of the call the option to cleanly terminate the INVITE transaction before the call is answered (e.g., hanging up the ringing call in the previous example).

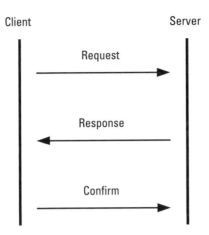

Figure 1.2   Three-way handshake.

BYE: Once a call is established using a SIP INVITE request, it will eventually come to an end (e.g., in a two-way voice call between two parties). The BYE primitive request allows either of the clients to terminate the INVITE-initiated session (e.g., hanging up a voice call once the conversation is completed). It should be remembered that the previously mentioned CANCEL primitive differs in that it is specifically for requesting to terminate a session that has not been established yet. The BYE request primitive is for terminating established (connected) INVITE-initiated interactions.

REGISTER: For SIP requests such as INVITE to reach a user, a dynamic mechanism is required that enables a client to inform a server that it is on the network and available for communication. With such location information, a trusted host server can then receive SIP requests on behalf of a client and direct it to the appropriate place. This can then apply to multiple devices, using a single SIP identifier for applications such as "find-me." A client can then also remove the location and routing information from the server (e.g., turning off a mobile handset, which might result in voicemail being activated instead). This server role is as a registrar who collects important information that is then used by a location service to find users. The SIP REGISTER primitive is used by the client to convey its current location information.

OPTIONS: While SIP is a successful multimedia negotiation protocol, it is sometimes useful to be able to obtain information about a potential or existing interaction before you attempt an operation. The OPTIONS primitive allows for the probing of both servers and endpoints to obtain important information such as what SIP extensions are supported and possible media types that are supported.

That introduces the basic core primitives involved in SIP and is only half the story for a request/response protocol. SIP also defines a series of response messages that are used in conjunction with the previously defined requests. The following represents the ranges of response codes used in SIP (for more information on specific response codes, see RFC 3261):

- *1xx: SIP Provisional Response*—A SIP response in the range of 100 to 199 is a provisional response indicating that a request has been received and is being processed.

- *2xx: SIP Success Response*—A SIP response in the range of 200 to 299 is a success response indicating that the request was received and has been accepted for a multimedia session.

- *3xx: SIP Redirection Response*—A SIP response in the range 300 to 399 is a redirection response indicating that request was understood and that the client needs to contact an alternative location to complete the multimedia session.

- *4xx: SIP Client Error*—A SIP response in the range 400 to 499 is a failure response indicating that the receiving entity could not complete the multimedia session for some reason (e.g., invalid syntax).

- *5xx: SIP Server Error*—A SIP response in the range 500 to 599 is a failure response indicating that the receiving server could not complete the valid multimedia session due to a problem occurring.

- *6xx: SIP Global Failure*—A SIP response in the range 600 to 699 is a failure indicating that the receiving entity, and any other server, could not complete the valid multimedia session for some reason (e.g., the user may no longer exist in the system).

The most common request/response interaction within the SIP protocol focuses on the INVITE primitive, which is used to initiate multimedia sessions. The following example provides a high-level view of the basic SIP INVITE interaction for a call setup and termination. As mentioned previously, for more detailed information on the SIP protocol, the reader is advised to consult the previously referenced specialist texts.

Figure 1.3 provides an illustrative depiction of a simple, basic SIP INVITE interaction for the purposes of establishing and terminating a multimedia session between two clients.

The SIP interaction starts when Chris decides to call Kristoffer from his SIP-enabled device. An INVITE request is created and sent to the local domain SIP proxy server, as depicted by (1) in Figure 1.3. It represents a session request from Chris to Kristoffer. The actual request is shown in Figure 1.4. (Note: The SDP payload has been left out for simplicity.)

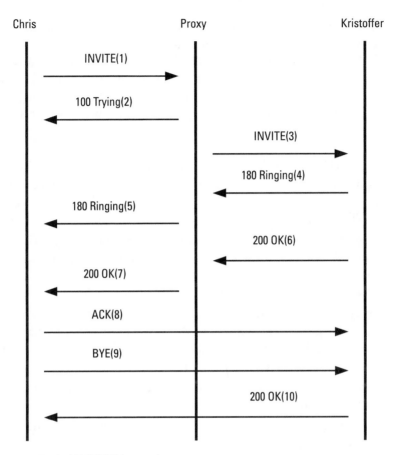

Figure 1.3 Basic SIP INVITE interaction.

```
INVITE sip:kristoffer@sipservlet_example.com SIP/2.0
Via: SIP/2.0/UDP sipservlet_example.com;branch=z9hG483JKSJ8ew9
Max-Forwards: 70
To: Stoffe < kristoffer@sipservlet_example.com >
From: Chris <chris@sipservlet_example.com >;tag=8327489874
Call-ID: fj8493ijf984ulw94@sipservlet_example.com
CSeq: 1 INVITE
Contact: <sip:chris@pc.sipservlet_example.com >
Content-Type: application/sdp
Content-Length: 150

(Chris's SDP is not shown.)
```

Figure 1.4 SIP INVITE request.

A key thing to notice in Figure 1.4 is the first line of the example, which determines the request type and the destination, known as the Request URI (or R-URI). The R-URI has a value of "INVITE" to indicate the appropriate request primitive type followed by a SIP URI indicating the destination of the request. The following lists the remaining headers from Figure 1.4 and their high-level meaning (note: these definitions apply to the remaining messages in this example as well):

Via: Provides information of the path that has been traversed by a SIP request. Each entity that forwards a SIP request must add its own "Via" header, which acts like a stamp of visitation. The list of SIP "Via" headers acts as trace route path for responses that must traverse the direction opposite to requests.

Max-Forwards: Contains a number indicating the maximum number of hops a SIP request can visit before it becomes invalid. This prevents requests from looping infinitely in the system. Each node that processes a SIP request should decrement the value of the Max-Forwards header.

To: The "To" SIP header contains the original recipient of the SIP request. Unlike the previously described R-URI, which is a dynamic representation of a requests destination, it does not change and allows the receiving entity to view the original recipient of the request.

From: Similar semantics as SIP "To" header, except it indicates the originator of the SIP request.

Call-ID: Used as a unique token for identifying a particular series of SIP requests.

CSeq: Within a series of SIP requests (as defined for the SIP "Call-ID" header) the CSeq provides a numerically increasing value within the scope of a SIP "Call-ID" header for messaging ordering purposes.

Contact: The SIP "Contact" header in this context indicates the physical location at which the SIP client can be reached, for subsequent messages to be exchanged within a series of related request/response interactions.

Content-Type: The SIP "Content-Type" header provides the media type that is being used in the attached payload of the SIP request.

Content-Length: The SIP "Content-Length" header provides the length of the attached payload of the SIP request. The combination of "Content-Type" and "Content-Length" allows a SIP parser to extract the exact payload with the correct multimedia session context.

The proxy server will receive the INVITE request from Chris and respond with a "SIP 100 Trying" response to signify that the request has been received and is being processed, which is indicated by (2) in Figure 1.3 and shown in Figure 1.5.

```
SIP/2.0 100 Trying
Via: SIP/2.0/UDP sipservlet_example.com;branch=z9hG483JKSJ8ew9;
received=192.0.2.10
To: Stoffe < kristoffer@sipservlet_example.com >
From: Chris <chris@sipservlet_example.com >;tag=8327489874
Call-ID: fj8493ijf984ulw94@sipservlet_example.com
CSeq: 1 INVITE
Content-Length: 0
```

Figure 1.5 "100 Trying."

The proxy server will then look to see if it has information relating to the location of Kristoffer, which could have been added either using the previously described SIP REGISTER message or using some other third-party means. On successfully retrieving the location information for Kristoffer, the SIP proxy forwards the request to his client, as illustrated by (3) in Figure 1.4 and shown in Figure 1.6.

Note that the proxy has inserted itself into the INVITE request by adding a SIP "Via" header and has also altered the R-URI to indicate the physical location of Kristoffer [as opposed to the Address of Record (AOR), which is a public domain-level identifier used in a similar manner to e-mail addresses]. Kristoffer's SIP client will receive and process the INVITE request and alert him that he has

```
INVITE sip:kristoffer@pc.sipservlet_example.com SIP/2.0
Via: SIP/2.0/UDP proxy.sipservlet_example.com;branch=z9hG48jHks7ds
Via: SIP/2.0/UDP sipservlet_example.com;branch=z9hG483JKSJ8ew9
Max-Forwards: 70
To: Stoffe < kristoffer@sipservlet_example.com >
From: Chris <chris@sipservlet_example.com >;tag=8327489874
Call-ID: fj8493ijf984ulw94@sipservlet_example.com
CSeq: 1 INVITE
Contact: <sip:chris@pc.sipservlet_example.com >
Content-Type: application/sdp
Content-Length: 150

(Chris's SDP is not shown.)
```

Figure 1.6 Proxy SIP INVITE.

an incoming request. The SIP client will also send out a SIP provisional response to indicate that Kristoffer has been alerted. This is illustrated in Figure 1.4 by (4), which shows a SIP "180 Ringing" response (in Figure 1.7) being sent back to the proxy server.

Kristoffer's SIP client used the SIP "Via" header inserted in the INVITE request it received to route the SIP response to the correct location. On receiving the SIP "180 Ringing" response, the proxy server will remove the SIP "Via" header that it inserted and forward the request onward to Chris's SIP client, as illustrated by (5) in Figure 1.4 and shown in Figure 1.8.

Again, the routing information for sending the response to the correct location is obtained from the SIP "Via" header previously inserted by Chris's SIP client. Eventually Kristoffer answers the request at his SIP client. This results in a SIP "200 OK" response being generated and sent in a similar manner to the previous SIP "180" response. The SIP "200 OK" response is sent from Kristoffer's SIP client, as illustrated by (6) in Figure 1.4 and shown in Figure 1.9.

The SIP "200 OK" message is then sent from the proxy server to Chris's SIP client, as illustrated by (7) in Figure 1.4 and shown in Figure 1.10.

```
SIP/2.0 180 Ringing
Via: SIP/2.0/UDP proxy.sipservlet_example.com;branch=z9hG48jHks7ds
Via: SIP/2.0/UDP sipservlet_example.com;branch=z9hG483JKSJ8ew9
To: Stoffe < kristoffer@sipservlet_example.com > tag=890092834
From: Chris <chris@sipservlet_example.com >;tag=8327489874
Call-ID: fj8493ijf984ulw94@sipservlet_example.com
CSeq: 1 INVITE
Contact: <sip:kristoffer@pc.sipservlet_example.com >
Content-Length: 0
```

Figure 1.7 SIP "180 Ringing" from client.

```
SIP/2.0 180 Ringing
Via: SIP/2.0/UDP sipservlet_example.com;branch=z9hG483JKSJ8ew9
To: Stoffe < kristoffer@sipservlet_example.com > tag=890092834
From: Chris <chris@sipservlet_example.com >;tag=8327489874
Call-ID: fj8493ijf984ulw94@sipservlet_example.com
CSeq: 1 INVITE
Contact: <sip:kristoffer@pc.sipservlet_example.com >
Content-Length: 0
```

Figure 1.8 SIP "180 Ringing" from proxy.

```
SIP/2.0 200 OK
Via: SIP/2.0/UDP proxy.sipservlet_example.com;branch=z9hG48jHks7ds
Via: SIP/2.0/UDP sipservlet_example.com;branch=z9hG483JKSJ8ew9
To: Stoffe < kristoffer@sipservlet_example.com > tag=890092834
From: Chris <chris@sipservlet_example.com >;tag=8327489874
Call-ID: fj8493ijf984ulw94@sipservlet_example.com
CSeq: 1 INVITE
Contact: <sip:kristoffer@pc.sipservlet_example.com >
Content-Length: 150
```

(Chris's SDP is not shown)

Figure 1.9 SIP "200 OK" from client.

On receiving the SIP "200 OK" response, Chris's SIP client completes the three-way INVITE handshake by generating a SIP acknowledgment protocol message (ACK). The SIP ACK message is sent directly to Kristoffer's SIP client and signifies the completion of the multimedia session setup. The ACK message is illustrated by (8) in Figure 1.4 and represented in Figure 1.11. It should be remembered that the ACK request forms part of a three-way handshake as part of a SIP INVITE interaction and does not generate a SIP response. The ACK request is sent directly to Kristoffer's SIP client using the direct information provided in the SIP "Contact" header that was present in the "200 OK" response. Note that this value has now been substituted into the R-URI of the ACK message in Figure 1.11. Once the initial SIP interaction takes place, SIP messages are sent directly using the values populated in the SIP "Contact" header.

```
SIP/2.0 200 OK
Via: SIP/2.0/UDP sipservlet_example.com;branch=z9hG483JKSJ8ew9
To: Stoffe < kristoffer@sipservlet_example.com > tag=890092834
From: Chris <chris@sipservlet_example.com >;tag=8327489874
Call-ID: fj8493ijf984ulw94@sipservlet_example.com
CSeq: 1 INVITE
Contact: <sip:kristoffer@pc.sipservlet_example.com >
Content-Length: 150
```

(Chris's SDP is not shown)

Figure 1.10 SIP "200 OK" from proxy.

```
ACK sip:kristoffer@pc.sipservlet_example.com SIP/2.0
Via: SIP/2.0/UDP sipservlet_example.com;branch=z9hG4hd73HUI
Max-Forwards: 70
To: Stoffe < kristoffer@sipservlet_example.com > tag=890092834
From: Chris <chris@sipservlet_example.com >;tag=8327489874
Call-ID: fj8493ijf984ulw94@sipservlet_example.com
CSeq: 1 ACK
Content-Length: 0
```

Figure 1.11   SIP "ACK" from client.

At this point, Chris and Kristoffer are exchanging media (e.g., voice/video/IM). At some point in the future, Chris decides to stop the session and terminates on his SIP client. This results in a SIP "BYE" message being generated and sent to Kristoffer's SIP client directly. This is illustrated by (9) in Figure 1.4 and shown in Figure 1.12.

On receiving the SIP "BYE" request, Kristoffer's SIP client responds with a SIP "200 OK" response to complete the transaction. This is illustrated by (10) in Figure 1.4 and shown in Figure 1.13.

```
BYE sip:kristoffer@pc.sipservlet_example.com SIP/2.0
Via: SIP/2.0/UDP sipservlet_example.com;branch=z9hGmHas7hj
Max-Forwards: 70
To: Stoffe < kristoffer@sipservlet_example.com > tag=890092834
From: Chris <chris@sipservlet_example.com >;tag=8327489874
Call-ID: fj8493ijf984ulw94@sipservlet_example.com
CSeq: 2 BYE
Content-Length: 0
```

Figure 1.12   SIP "BYE" from client.

```
SIP/2.0 200 OK
Via: SIP/2.0/UDP sipservlet_example.com;branch= z9hGmHas7hj
To: Stoffe < kristoffer@sipservlet_example.com > tag=890092834
From: Chris <chris@sipservlet_example.com >;tag=8327489874
Call-ID: fj8493ijf984ulw94@sipservlet_example.com
CSeq: 2 BYE
Content-Length: 0
```

Figure 1.13   SIP "200 OK" from client.

This message signifies the completion of an INVITE-based multimedia session, and media between Chris and Kristoffer stops. This completes our basic introduction to the SIP protocol and one of its most important constructs.

1.2 SIP Servlets and the SIP Servlet Vision

After that brief introduction to the SIP protocol, it is now time to introduce SIP Servlet technology. Most readers will already be familiar with HTTP Servlet technology; for those who are not, a brief introduction is included later. SIP Servlets are very similar to HTTP Servlets—they share the same core Java package—but are related to their appropriate signaling protocols (SIP for SIP Servlets and HTTP for HTTP Servlets). While it can be stated that they have a very close relationship, and can even integrate extremely tightly within an application (HTTP and SIP convergence are discussed later in the book), the reader must also be made aware that, due to the asynchronous nature of the SIP protocol, either side can initiate a request within a potentially long-lived session. This imposes different requirements and demonstrates how SIP Servlets differ greatly from HTTP Servlets. The developer needs to think of issues such as:

- Storage and retrieval of application data for servicing subsequent and related requests;
- Subsequent requests arriving from either direction for a particular session;
- Dealing with a forked SIP request and potentially multiple positive responses;
- Deciding on the specific role that an application will assume.

Add on top of that the ability for any node, at any time, to originate requests unexpectedly, and the separation between SIP and HTTP Servlets becomes apparent. In short, the SIP Servlet vision is to provide an architecture that has the ability to fully integrate into Java Enterprise Edition (JEE) while providing similar functionality to HTTP Servlets (such as life-cycle management, abstracted Application Programming Interface (API), and so forth) and also fully utilizing the power of the SIP protocol and its extensions for next-generation multimedia communications. Now let's have a brief look at the major components of the SIP Servlet architecture: the SIP Servlet container and the SIP Servlet.

JEE has the concept of specific containers that are hosts to varying technologies in the Architecture. Examples, which are briefly introduced later in this section, include HTTP Servlets, which are hosted in a Web container, and Enterprise JavaBeans (EJBs), which are hosted in an EJB container. Containers are part of a larger overall JEE Application Server that is responsible for managing

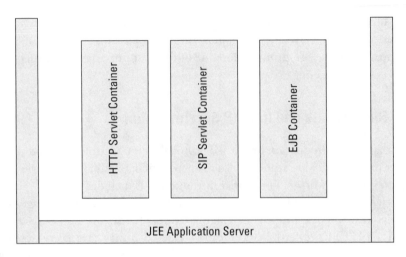

Figure 1.14 JEE container architecture.

such containers. See Figure 1.14 for an example of multiple containers being hosted within a JEE Application Server.

It should be noted that, while this is expected to be the most common usage of SIP Servlet containers moving forward, there is no requirement for SIP Servlets containers to be deployed within a JEE Application Server. In fact, it is extremely feasible to have stand-alone instances of SIP Servlets containers that have no dependency on JEE (e.g., the SIP Servlet container could be deployed using just a SIP stack's native API).

The SIP Servlet container has a number of responsibilities when deployed, primarily focusing on:

- Deployment and management of SIP Servlets-based applications;
- Security related to SIP Servlet applications;
- Management of SIP protocol-level functionality, including listening ports, interfaces, protocol abstraction, and protocol policing.

Hosted on SIP Servlet containers are SIP Servlet application archives. In a manner similar to the way Web-based applications are bundled in custom deployment units, called Web archives, which use the ".war" file extension, and JEE applications are bundled using the ".ear" file extension, SIP Servlet archives are bundled using the ".sar" file extension. For example, a SIP Servlet application called "FindMe" would be packaged under the file "FindMe.sar." Servlet containers generally manage a SIP Servlet application and interact using a common API and callback mechanism that enables the container to act on the SIP Servlet's behalf from a signaling perspective. Due to the open-standards approach of SIP Servlets' architecture, using a common API and structured packaging, a SIP Servlet application is able to run on any compliant container. This promotes an open marketplace for application and container developers. This, in turn, encourages competition and increases the quality and adoption rate of the technology. It also improves choice for technology purchasers, who are not locked into a specific container product or application producer. A more detailed overview of SIP Servlet applications is included in Chapter 3.

1.3 Java Enterprise Edition

SIP Servlet 1.1 technology (JSR 289 [12]) is being developed by the Java Community Process (JCP). The JCP has defined a standard architecture called Java Enterprise Edition (JEE), which specifies a standard platform for hosting JEE-compliant applications. The ultimate goal of SIP Servlet technology is to be adopted as part of the official JEE standard for enterprise-level application servers. This section will not provide a tutorial of JEE and its associated technologies but has been included as an introduction to related technologies. While SIP Servlets do not rely on any specific JEE technology, and in fact have been deployed stand-alone since their conception, it is important to recognize their relationship with other JEE technologies. The true power and flexibility of SIP Servlets becomes evident when used as an enabler for a wide variety of applications and deployments. First, we will take a closer look at the JEE Servlet specification that SIP Servlets derive from and then introduce the two extremely important technologies of annotations and Enterprise JavaBeans.

1.3.1 Servlet Specification

The Servlet specification specifies how to package and deploy a Web application as either stand-alone or part of a larger JEE application. A Servlet is a deployment unit (Web archive, or .war file) that complies with the associated Servlet specification [13]. Application servers wishing to host Servlets must have a compliant Servlet container for hosting Web archive applications. The containers are extensions to core Web technology provided by industry standard protocols, such as the Hypertext Transfer Protocol, and use the classic request/response protocol interactions to provide advanced business logic and dynamic decision making. On deploying a Servlet, the container then manages the life cycle of the Web application from initialization to destruction. Figure 1.15 illustrates a simplistic view of HTTP Servlets that have been deployed in a Servlet container.

As mentioned previously in this section, the SIP protocol was modeled to some extent on the HTTP protocol. For this reason, it was a natural fit for SIP Servlets to follow an extremely similar model to HTTP Servlets. If designed

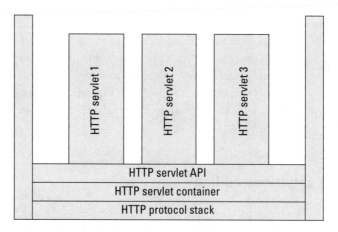

Figure 1.15 HTTP Servlet container.

appropriately, it was viewed that a tight relationship between HTTP and SIP Servlets would lead to a more powerful programming paradigm. The Servlet specification defines a generic part of its definition that is not HTTP specific and is bundled in the Java package "javax.servlet." The specific HTTP part of the package for HTTP Servlets is, then, defined under the package name "javax.servlet.http." In a similar manner, SIP Servlets build on the generic Servlet API under the package name "javax.servlet.sip," as shown in Figure 1.16.

As all HTTP Servlet containers have to support both the generic Servlet package and the HTTP package, in a similar way, a SIP Servlet container has to support both the generic Servlet package and the SIP Servlet package. Introduced in the SIP Servlet specification is also the concept of convergence between HTTP and SIP Servlet containers. This allows an HTTP Servlet-based application to interact with a SIP Servlet-based application and vice versa. Convergence of the Web and JEE world with SIP provides a powerful and flexible application environment for future communication evolution. The concept of SIP Servlets converging with HTTP Servlets and other JEE technologies will be discussed in extensive detail in later chapters of the book.

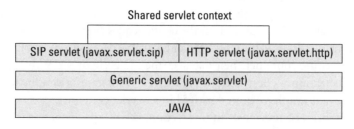

Figure 1.16 Package structure.

1.3.2 Annotations

The release of Version 5 of JEE included a number of new and important technologies for improving the usability and ease of development in JEE-compliant
application servers. One such technology, defined in JSR 175 [14], specifies an
annotation function for the Java language. The ability to annotate classes, methods, and fields provides a flexible programming model that can be used to inject
various resource types into an application and supports a move away from traditional deployment descriptors that appear in many JEE technologies. As SIP
Servlets make use of both deployment descriptors and the injection of various
resources into applications, Java annotations play an increasingly important role
in SIP Servlet 1.1 technology. More specific information on SIP Servlet-specific
annotations is given in Chapter 3.

Probably the most practical annotation added in SIP Servlet specification
is the @SipServlet annotation. If there is only a single Servlet in a SIP application, then nothing more than an empty "sip.xml" descriptor is needed. A simplest possible application would look as follows:

```
@javax.servlet.sip.annotation.SipServlet
public class SimplestServlet extends javax.servlet.sip.SipServlet {
  @Override
  protected void doMessage(SipServletRequest req) throws ServletException,
      IOException {
    SipServletResponse resp = req.createResponse(SipServletResponse.SC_OK);
    resp.send();
  }
}
```

1.3.3 Enterprise JavaBeans (EJB)

Enterprise JavaBeans is a well-established technology that has been evolving for
a number of years and, at the time of this writing, is in Version 3.0. The general
EJB architecture focuses on providing a solution for reusable, object-oriented
business logic components that can be linked to provide larger applications and
services. EJB components can be colocated or distributed across different network
locations using the transactional and scalable architecture.

To clarify the focus of EJB, the overall goals of the technology from the EJB
3.0 specifications [15] are included here:

- The Enterprise JavaBeans architecture will be the standard component
  architecture for building object-orientated business applications in the
  Java programming language.

- The Enterprise JavaBeans architecture will be the standard component architecture for building distributed business applications in the Java programming language.

- The Enterprise JavaBeans architecture will support the development, deployment, and use of web services.

- The Enterprise JavaBeans architecture will make it easy to write applications: Applications developers will not have to understand low-level transaction and state management details, multithreading, connection pooling, or other complex low-level APIs.

- Enterprise JavaBeans applications will follow the Write Once, Run Anywhere philosophy of the Java programming language. An enterprise bean can be developed once, and then deployed on multiple platforms without recompilation or source code modification.

- The Enterprise JavaBeans architecture will address the development, deployment and run time aspects of an enterprise application's life-cycle.

- The Enterprise JavaBeans architecture will define the contracts that enable tools from multiple vendors to develop and deploy components that can interoperate at run time.

- The Enterprise JavaBeans architecture will make it possible to build applications by combining components deployed using tools from different vendors.

- The Enterprise JavaBeans architecture will provide interoperability between Enterprise JavaBeans and Java Platform, Enterprise Edition (Java EE) components as well as non-java programming language applications.

- The Enterprise JavaBeans architecture will be compatible with existing server platforms. Vendors will be able to extend their existing products to support Enterprise JavaBeans.

- The Enterprise JavaBeans architecture will be compatible with other Java programming language APIs.

- The Enterprise JavaBeans architecture will be compatible with CORBA protocols.

The rules taken from the EJB 3.0 specification provide an indication of the importance of the technology to the general JEE architecture. EJB provides the core business component technology including how they are reused, managed, invoked, and secured. As SIP Servlet aims to be integrated in future versions of JEE technology, its relationship with EJB is key to improving the value proposition to both application developers and application deployers. More information will be provided later in the book on how SIP Servlets 1.1 facilitates both EJB and annotation integration, opening the door to the wider JEE architecture.

Entity bean EJBs can now be used inside a SIP application bundled inside an EAR (Enterprise Archive). An entity bean can store a Java Bean persistently in a Structured Query Language (SQL) database using the "EntityManager" interface. Also POJOs (Plain Old Java Objects) can become EJB entity beans with a call to the "EntityManager." This reduces the development time of applications that need to store data between server restarts.

Another important EJB bean is the session bean, which can be both stateful and stateless. This kind of bean can be used as a remote integration point for a SIP application. Imagine writing a presence or a chat server that should include a remote interface to add new users. It is easy to annotate a function that adds a user with the @Stateless annotation, making the function accessible over Remote Method Invocation (RMI) (using a client EJB stub).

The third type of EJB bean is the Message Driven Bean (MDB). The MDB is simply an asynchronous execution point. The most interesting scenario when MDBs are used in the context of the SIP container is in conjunction with the JEE connector framework. The connector has an outbound direction when calling from within the JEE application server. One common example for the outbound connection usage is to implement Java Database Connectivity (JDBC). In the case of the database, they are often request/response interactions. For asynchronous events on a connector, the MDB is used as an input queue. So, for dealing with asynchronous events, an MDB would be called and could, in turn, get hold of the SIP framework to generate a SIP message.

```
@PersistenceUnit(unitName="PuSample")
private EntityManagerFactory emf;
MyPojo pojo = new MyPojo("Bob",""sip:bob@sipservlet.net");
EntityManager em = emf.createEntityManager();
em.persist(pojo); //Will create an EJB and store it in the DB
```

References

[1]   Rosenberg, J., et al., "SIP: Session Initiation Protocol," RFC 3261, Internet Engineering Task Force, June 2002.

[2]   Handley, M., et al., "SIP: Session Initiation Protocol," RFC 2543, Internet Engineering Task Force, March 1999.

[3]   Rosenberg, J., and H. Schulzrinne, "Reliability of Provisional Responses," RFC 3262, Internet Engineering Task Force, June 2002.

[4]   Rosenberg, J., and H. Schulzrinne, "Session Initiation Protocol (SIP): Locating SIP Servers," RFC 3263, Internet Engineering Task Force, June 2002.

[5]   Rosenberg, J., and H. Schulzrinne, "An Offer/Answer Model with the Session Description Protocol (SDP)," RFC 3264, Internet Engineering Task Force, June 2002.

[6]   Handley, M., V. Jacobson, and C. Perkins, "SDP: Session Description Protocol," RFC 4566, Internet Engineering Task Force, July 2006.

[7]   Roach, A. B., "Session Initiation Protocol (SIP)—Specific Event Notification," RFC 3265, Internet Engineering Task Force, June 2002.

[8]   Rosenberg, J., "A Presence Event Package for the Session Initiation Protocol (SIP)," RFC 3856, Internet Engineering Task Force, August 2004.

[9]   Olson, S., G. Camarillo, and A. B. Roach, "Support for IPV6 in the Session Description Protocol (SDP)," RFC 3266, Internet Engineering Task Force, June 2002.

[10]   Johnston, A. B., *SIP: Understanding the Session Initiation Protocol,* 2nd ed., Norwood, MA: Artech House, 2003.

[11]   Fielding, R., et al., "Hypertext Transfer Protocol—HTTP/1.1," RFC 2616, Internet Engineering Task Force, June 1999.

[12]   SIP Servlet Specification, Version 1.1, JSR 289, Java Community Process, August 2008.

[13]   Java Servlet Specification, Version 2.4, JSR 154, Java Community Process, September 2007.

[14]   A Metadata Facility for the Java Programming Language, JSR 175, Java Community Process, September 2004.

[15]   Enterprise JavaBeans 3.0, JSR 220, Java Community Process, November 2007.

2

The SIP Servlet Container

The introduction provided a high-level overview of a SIP Servlet container and its core features and responsibilities. This section focuses at a much lower level of granularity relating to container functionality and its most important operations. A SIP Servlet container, or Servlet engine as it is sometimes known, provides a consistent architecture for running compliant home-engineered or third-party SIP-based applications. It is either deployed as part of a larger JEE application server, converged SIP, and HTTP container or as a stand-alone SIP entity.

2.1 Container Responsibilities

One of the primary goals of SIP Servlet technology has always been to parallel related technologies such as HTTP Servlets and EJB in providing application developers with a programming paradigm that abstracts as much complication away as possible. To achieve this, SIP Servlet containers have a number of core responsibilities that allow application developers to concentrate on important business logic rather than mundane protocol-related tasks.

2.1.1 Life-Cycle Management

As briefly mentioned in the introduction and covered in more detail later in the book, SIP Servlet applications are bundled into an appropriately structured archive that uses the ".sar" extension. On deployment of a SIP Servlet application to a container, a number of life-cycle management tasks have to be followed for successful deployment, running, and undeployment of an application. Figure 2.1

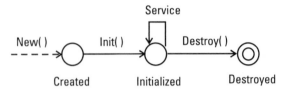

Figure 2.1   SIP Servlet Container Life Cycle.

is taken from the SIP Servlet 1.1 specification [1] and illustrates the life-cycle of a SIP Servlet application being deployed in a container.

Initially, the SIP Servlet container will inspect the application deployment unit (".sar" file) for the appropriate Java class representing the Servlet. It will also scan the class files and bundled Java archive (JAR) file for any SIP Servlet specific annotations. At this stage, the container is in the "initialize" state (represented by "init" in Figure 2.1) of the life cycle, which results in all appropriate configurations being passed into the application from the associated deployment descriptor file and any logic executed in the Servlets "init" method. Once initialization has completed, the life cycle is able to offer service (which means it is ready to be invoked by the container on receiving appropriate SIP signaling). At some stage in the future, the container (possibly as a result of user interaction) will deactivate the SIP Servlet application and remove it from active service, resulting in its not being able to receive any SIP signaling. Such an operation results in the "destroy" method of the application being called before it is from the list of active container-managed applications. This represents the full life cycle of a SIP Servlet-based application.

It should also be noted that, due to the varying roles (as will be discussed later in the book) that a SIP Servlet application can adopt in relation to SIP signaling, a number of problem areas and race conditions occur when deploying SIP Servlet applications. For example, an application can be triggered to generate SIP protocol messages without any incoming signaling. Returning to Figure 2.1, there is no way a Servlet application can know if the "init" method has totally been completed, and so it is not aware of exactly when it can initiate SIP signaling. This could result in SIP messages being generated and sent before all initialization tasks have taken place. To overcome this problem, SIP Servlet 1.1 has a specific listener that the container has to invoke only when it truly knows the SIP Servlet application has been fully initialized and is ready for service. Any application that is intending to create unrelated SIP signaling or even to carry out other tasks in the initialization phase must implement the listener to learn when it is able to carry out normal service.

SipServletListener.servletInitialized(SipServletContextEvent ce)

2.1.2 Protocol Compliance

A SIP Servlet container, along with the SIP Servlet applications, bears some of the burden of ensuring SIP protocol compliance by both hosted applications and interactions with external SIP entities. A container would naturally be responsible for certain objects that form part of the SIP Servlet API and are used by applications. These are described in SIP Servlet architecture as "container managed" and include well-known constructs such as application sessions, protocol sessions/states, and SIP URI objects (all of which will be discussed in more detail later). When a violation of any states occur, the container must act as the protocol police and ensure container-managed objects are not compromised, by throwing an appropriate Java exception as an indication to the application. A detailed list of appropriate exceptions and the conditions under which they are thrown is included in the SIP Servlet 1.1 specification.

2.1.3 Mapping Requests to Servlets

One of the most important jobs that a container has to carry out once it has established that a SIP Servlet application is in service is supplying it with appropriate SIP signaling. The invocation of hosted applications has certainly evolved during the technology's lifetime, and with Version 1.1 of SIP Servlets, we have a mechanism appropriate for advanced next-generation deployments.

Invocation of SIP Servlet applications has a hierarchical approach that is phased from the moment an initial SIP signaling request enters the container. The first stage involves the container's invoking a special container API called the Application Router (AR) interface, which is specified as part of the SIP Servlet programming API. An AR is a logical function that implements the AR API with the intention of receiving a request from the container in the up-call and supplying the SIP Servlet application to be visited next. This process is recursive until no more applications hosted on the container remain to be visited in the context of the SIP signaling request and it carries on its journey in the SIP network. Figure 2.2 illustrates how a container receives a request and consults the AR API for the purpose of receiving a SIP Servlet application to service the request.

The implementation of the logic behind the AR API interface is totally vendor specific and allows controllers of networks to deploy and select applications however they choose. As long as the common AR API interface is implemented, then the decision-making process can be as complex or as simple as required and can be in any programming or scripting language. SIP Servlets 1.1 does provide details in an appendix of a default application router, but this is just to provide a basic example of how an AR could work. The Default Application Router (DAR) is provided in order for SIP Container providers to be able to test and verify their

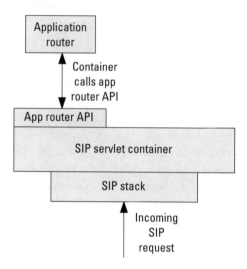

Figure 2.2 Basic Application Router interaction.

implementation. At the same time, the TCK (Test Compatibility Kit) also provides a Reference Implementation (RI) in order for the Application Router developers to test and validate their implementation of their deployment-specific AR. An extensive, dedicated section on application composition and routing follows in Chapter 4.

Once the AR has been consulted and an application name has been returned for a SIP signaling request, it is the container's responsibility to dispatch the SIP message to the appropriate application archive (".sar" file). A SIP Servlet application is composed of a deployment descriptor file, which provides details relating to the application, and any number of SIP Servlet class files, which are all packaged into the compliant structure (see later section for more detail relating to packaging). See Figure 2.3 for a simple view of the SIP Servlet application (".sar" file).

A container must determine which of the SIP Servlet class files should be visited within SIP Servlet application. SIP Servlet 1.0, the previous version of the specification, provided dedicated XML-based filter-mapping functions so that containers could take a request that has been dispatched to a SIP Servlet application and then direct it to the appropriate Servlet class file. This mechanism was known as Servlet Mappings. SIP Servlet 1.1 has defined a replacement for Servlet mappings based on the concept that every incoming SIP message will automatically be sent by the container to a default Servlet class—named the "main Servlet." If it is decided that the request should be serviced by another Servlet class within the application, it is programmatically dispatched. Both of these container selection mechanisms remain valid in SIP Servlet 1.1, with the latter, main Servlet class recommended. The following sections will provide detail of both mechanisms.

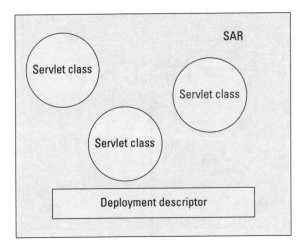

Figure 2.3 Servlet archive (.sar) representation.

2.1.3.1   Default Servlet (SIP Servlet 1.1)

Using this mechanism is relatively straightforward for the container, as it simply needs to identify the "main" Servlet in the application and overload the message. Only a single Servlet class in the application that contains multiple Servlets can be marked as a "main" Servlet and marking it is optional in applications that only contain a single Servlet. A Servlet marks itself as being the "main" Servlet by either including the Java annotation "@SipApplication mainServlet" element or adding the "<main-servlet>" element in the deployment descriptor for the application. Figure 2.4 illustrates how, once an application is selected using the Application Router, it can be dispatched by the container using the "main" Servlet mechanism.

The main Servlet will receive all SIP-related requests directed at the application, and so it must identify any requests that need to be serviced by an alternative Servlet and dispatch them appropriately. This is achieved using the "Request-Dispatcher" interface, which allows a programmer to pass request objects and associated properties between Servlet applications. Once the initial request has been dispatched to the appropriate Servlet using the "RequestDispatcher" interface, a developer is also able to configure the application so that all subsequent associated requests and responses are automatically passed to the correct Servlet and not the "main." This is achieved using the "setHandler" method that appears on in the SIP Servlet API (see later section for more information).

When the main Servlet is dispatching a request to another Servlet within the same application, it can then call "setAttribute" method on the "SipServlet Message." In this way it can pass parameter to the next receiving Servlet that should take care of the execution of the message. The receiving Servlet would call

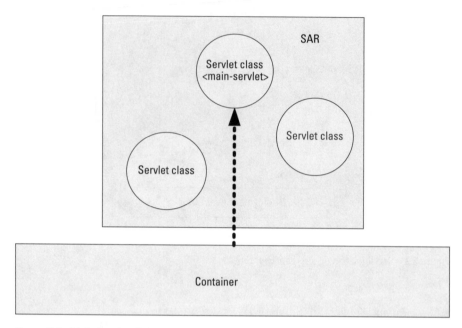

Figure 2.4 Main Servlet dispatch.

"getAttribute" method on the message to retrieve any values that the main Servlet might have stored for it. Attributes are visible only within one application, so they cannot be used as a utility in a composition of many different SIP applications.

2.1.3.2   Servlet Mappings (SIP Servlet 1.0)

Servlet mappings proved to be a very popular and well-known selection mechanism in SIP Servlet 1.0. While it has been replaced by the previously described "main" Servlet mechanism for selecting from multiple Servlets, it is still valid. The Servlet-mappings approach involves XML elements that appear in the deployment descriptor and tell the container explicitly which Servlet should be invoked for the SIP request. Figure 2.5 shows an example of a simple Servlet-mapping.

This very simple Servlet-mapping represents two Servlets in the same application. The first in the list instructs the container to dispatch any initial SIP INVITE requests to a Servlet class called "INVITEServlet." It also instructs the container to dispatch any initial SIP SUBSCRIBE requests to a Servlet class called "SUB-SCRIBEServlet." If a number of entries in the Servlet-mappings XML element match, then the first in the list is selected. While this is a well-known paradigm for selecting a Servlet Class within an application, it is fairly static and does not allow the same flexibility as the programmatically based "main" Servlet approach.

Note that only one of the two mechanisms must be used when using multiple Servlets within an application. Specifying both is an error, and the application should be rejected by the container when deployed.

```
<servlet-mapping>
        <servlet-name>INVITEServlet<servlet-name/>
        <pattern>
                <equal>
                        <var>request.method</var>
                        <value>INVITE</value>
                </equal>
        </pattern>

        <servlet-name>SUBSCRIBEServlet<servlet-name/>
        <pattern>
                <equal>
                        <var>request.method</var>
                        <value>SUBSCRIBE</value>
                </equal>
        </pattern>
</servlet-mapping>
```

Figure 2.5 Servlet mapping.

Note Servlet mapping is also supported in SIP Servlet 1.1, but the format has changed! This can cause some confusion on deployment. In SIP Servlet 1.0, the "<servlet-mapping>" tag is directly under the "<sip-app>" element. In SIP Servlet 1.1, the hierarchy is "<sip-app>, <servlet-selection>, <servlet-mapping>."

2.1.4 Receiving SIP Requests

As mentioned previously, the SIP Servlet architecture extends the core Servlet specification in a similar way to that of the HTTP Servlet. The core Servlet specification defines a "service" method on its interface that is the primary vehicle for sending and receiving messages. On receiving a message, the "service" interface is called in the relevant context, either for a message that is a request or for a response. Figure 2.6 shows the basic "service" method that appears on the core Servlet interface.

```
void service(ServletRequest req, ServletResponse res)
      throws ServletException, java.io.IOException
```

Figure 2.6 Service interface.

Using this "service" method as a basis when applied to the SIP Servlet API, the container invokes a SIP specific type of the "service" method. For a SIP signaling request, the "SIPServletRequest" interface is invoked, and for responses, the "SIPServletResponse" interface is invoked. Each time, the request and response objects are passed to the SIP Servlet API for processing, as shown in Figure 2.7.

It was found that, to make SIP Servlet programming more user-friendly, a lower level of granularity is required to reduce complexity when an application receives a request that has been dispatched by the container. While it is fine to use just the main SIP Servlet interface and receive SIP signaling in its basic request and response form, this leads to application developers' then having to parse SIP messages to obtain the type of message primitive (e.g., INVITE or BYE). It is for this reason that the SIP Servlet API defines subclass methods that are called directly from the "doRequest" method previously defined for receiving SIP requests. The new methods are representative of a list of primary SIP primitives that are used in applications:

- *doInvite* is invoked by the SIP Servlet "doRequest" method on receiving a SIP INVITE primitive, as defined in RFC 3261 [2].

- *doAck* is invoked by the SIP Servlet "doRequest" method on receiving a SIP ACK primitive, as defined in RFC 3261 [2].

- *doOptions* is invoked by the SIP Servlet "doRequest" method on receiving a SIP OPTIONS primitive, as defined in RFC 3261 [2].

- *doBye* is invoked by the SIP Servlet "doRequest" method on receiving a SIP BYE primitive, as defined in RFC 3261 [2].

- *doCancel* is invoked by the SIP Servlet "doRequest" method on receiving a SIP CANCEL primitive, as defined in RFC 3261 [2].

- *doRegister* is invoked by the SIP Servlet "doRequest" method on receiving a SIP REGISTER primitive, as defined in RFC 3261 [2].

- *doPrack* is invoked by the SIP Servlet "doRequest" method on receiving a SIP PRACK primitive, as defined in RFC 3262 [3].

- *doSubscribe* is invoked by the SIP Servlet "doRequest" method on receiving a SIP SUBSCRIBE primitive, as defined in RFC 3265 [4].

```
protected void doRequest(SipServletRequest req);
protected void doResponse(SipServletResponse req);
```

Figure 2.7 SIP Servlet implementation of service.

- *doNotify* is invoked by the SIP Servlet "doRequest" method on receiving a SIP NOTIFY primitive, as defined in RFC 3265 [4].

- *doMessage* is invoked by the SIP Servlet "doRequest" method on receiving a SIP MESSAGE primitive, as defined in RFC 3428 [5].

- *doInfo* is invoked by the SIP Servlet "doRequest" method on receiving a SIP INFO primitive, as defined in RFC 2976 [6].

- *doUpdate* is invoked by the SIP Servlet "doRequest" method on receiving a SIP UPDATE primitive, as defined in RFC 3311 [7].

- *doRefer* is invoked by the SIP Servlet "doRequest" method on receiving a SIP REFER primitive, as defined in RFC 3515 [8].

- *doPublish* is invoked by the SIP Servlet "doRequest" method on receiving a SIP PUBLISH primitive, as defined in RFC 3903 [9].

It is an extensive set of potential methods that can be used in SIP Servlet-based applications as the foundation for filtering incoming requests. Figure 2.8 illustrates an incoming request being dispatched by the container to the appropriate service method using the SIP Request-URI of a SIP request included in the figure (note that the rest of the SIP message is left out for the sake of simplicity).

The SIP Servlet API also allows total flexibility for new SIP methods that are either not covered or are proprietary. This can be achieved by overriding the base SIP Servlet "doRequest" method and using "super."

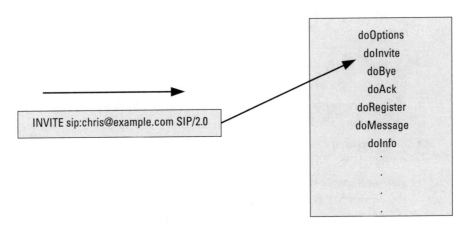

Figure 2.8 Dispatching of request.

```
protected void doRequest(SipServletRequest request)
       throws ServletException, IOException

{

       if (NewRequest.equals(request.getMethod())
              {
                     doNewRequest;
              } else
              {
                     Super.doRequest(request);
              }
}
```

Figure 2.9 Dispatching of unspecified request.

Figure 2.9 provides an example of "doRequest" being overridden and calling a new method (doNewRequest) if the method extracted for the incoming SIP request is equal to the value stored in the parameter "NewRequest." If it does not match, then the "else" block in the example is called and the request is dispatched appropriately.

2.1.5   Receiving SIP Responses

The handling of SIP Servlet responses when being dispatched to the SIP Servlet application by the container is very similar to the handling mechanism for requests. Whereas, for requests, the "doRequest" method dispatches to the appropriate SIP primitive method (as described in Figure 2.8), the "doResponse" method does the same when receiving a SIP response message. Depending on the status code of the SIP message, the response will be dispatched as follows:

- *doProvisionalResponse* is invoked by the SIP Servlet "doResponse" method on receiving a SIP response code in the range 101 to 199.
- *doSuccessResponse* is invoked by the SIP Servlet "doResponse" method on receiving a SIP response code in the range 200 to 299.
- *doRedirectResponse* is invoked by the SIP Servlet "doResponse" method on receiving a SIP response code in the range 300 to 399.
- *doErrorResponse* is invoked by the SIP Servlet "doResponse" method on receiving a SIP response code in the range 400 to 699.
- *doBranchResponse* is invoked once for every SIP branch in a forking proxy receiving a response. When all branches are completed then the "doResponse" is called only with the best response.

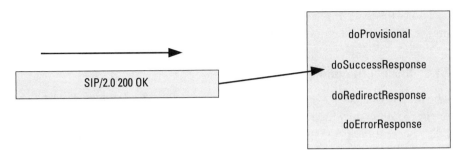

Figure 2.10 Dispatching of response.

Figure 2.10 illustrates an incoming response being dispatched by the container to the appropriate service method using the first line of a SIP response included in the figure (note that the rest of the SIP message is left out for the sake of simplicity).

We have now followed the initial stages of a SIP message arriving at a SIP Servlet container and being matched to the appropriate SIP application archive (deployment unit using the ".sar" file extension) and then matched to the appropriate SIP Servlet within the application archive by using either Servlet-mapping rules from the deployment descriptor or the "main" Servlet approach.

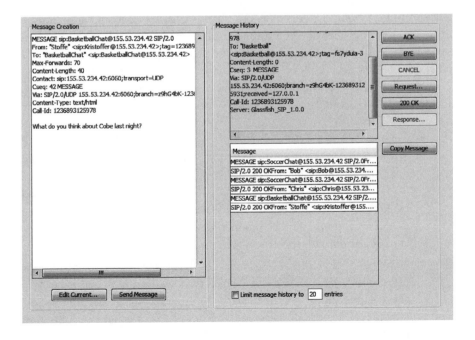

A SIP Servlet application is now able to apply appropriate application logic on the SIP protocol requests and responses.

2.1.6 Session Targeting

The SIP Servlet API is used for a wide variety of applications that have vastly different requirements. One of the most important usages is focused on multiparty calls where multiple user interactions are linked together, such as a conference call. Discussed in more detail later in this book are the general concepts surrounding how an application springs into life on receiving a new SIP request. In short, an umbrella object called an SIP Application Session is created to be responsible for storing application-specific data and correlating individual protocol sessions, called SIP Sessions. A quick example: On receiving a SIP INVITE request from user A, the container will dispatch to Application X using the mechanisms described in this section. On entering Application X, a new Application Session is created along with an appropriate SIP Session. The newly created Application Session could be, for example, representing a conference. If user B now sends a SIP INVITE request to the container, it can reach the same application using the mechanisms defined in this chapter, but this will result in another new Application Session and SIP Session being created. As a result, we have two independent Application Session and SIP Session instances that are in no way related and that would need some form of proprietary linkage mechanism to be associated.

Using the previous example, it becomes obvious that a container mechanism that allows incoming SIP requests to be correlated to an existing Application Session instance would be extremely useful and powerful. It would also increase efficiency as the number of Application Session objects whose life cycles the container has to create and manage reduces. The need for proprietary linkage mechanisms disappears, which ultimately encourages innovation and interoperability.

The SIP Servlet 1.1 architecture includes three mechanisms for associating new SIP signaling requests to existing Application Sessions:

- *Encode URI mechanism*—Involves creating a unique key using the SIP Servlet API (SipApplicationSession.encodeURI) method, which is then distributed to all potential users (distribution of such a key is out of scope of the architecture and therefore this book but could be accomplished by, for example, e-mail or Instant Message). If an encoded URI appears in a new SIP request, the container identifies it and makes the appropriate association with the existing Application Session.

- *Session Key mechanism*—The preferred approach to associating new SIP requests with an existing Application Session, it makes use of Java annotations and user-defined algorithms to decide if association should take place. A more detailed explanation of this mechanism follows in this section.

- *Targeted mechanism*—This is an optional-to-implement mechanism that is used in association with certain SIP extensions names, such as Join [10] and Replaces [11]. A more detailed explanation of this mechanism is also provided later in this chapter.

The Encode URI mechanism has been depreciated in SIP Servlet 1.1, because implementation experience proved the concept to be inadequate. It also does not fit with the new direction of SIP Servlet technology, which is very much focused on application composition chains rather than bypassing the selection process with a container dispatching a request directly to a SIP Servlet application. The remainder of this chapter will focus specifically on the remaining two mechanisms, session key and targeted, which very much fit into SIP Servlet 1.1 main concepts of application selection and chaining.

2.1.6.1 Session Key

The session key mechanism is very different from the session-targeted mechanism, which will be covered next. They both occur at different times in the application selection process and are therefore complementary, rather than overlapping or competing, functions. Earlier in this book, we looked at the traversal of an incoming request on its journey into the container and eventually into a SIP Servlet application. The session key mechanism applies at the stage in the selection process where the Application Router has already selected a SIP Servlet application. The container should then dispatch the request to the application, but instead of just automatically creating a newly associated Application Session, it checks using the session key mechanism to see if it should associate the newly created protocol session (SIP Session) with an existing Application Session.

The session key mechanism makes use of Java annotations, which were introduced briefly at the beginning of the book and are discussed in more detail later on. A new Java annotation called @SipApplicationKey is defined in the specification and should appear in a SIP Servlet application wishing to use this mechanism. The annotation takes a SIP request as input and returns a string as a key to represent the Application Session created as a result. According to the specification, this method needs to be a public and static one. The following is a simple example of how the @SipApplicationKey would appear in one of the classes contained in the SIP Servlet application:

```
@javax.servlet.sip.annotation.SipServlet

public class SipKeyServlet extends SipServlet {
  @Override
  protected void doMessage(SipServletRequest req) throws ServletEx-
ception, java.io.IOException {
    log("Received message : "+req.getContent().toString()+" in SAS
with id = "+req.getApplicationSession().getId());
```

```
    req.createResponse(200).send();
}


@SipApplicationKey

public static String getKey(SipServletRequest req) {
    SipURI source = (SipURI) req.getFrom().getURI();
    SipURI target = (SipURI) req.getTo().getURI();
    String confName = target.getUser();
    System.out.println("Assigning user " + source.getUser() + " to
SipApplicationSession with key = "+confName);
    return confName;
  }
}
```

```
MESSAGE sip:SoccerChat@147.214.199.3 SIP/2.0
From: "Alice"<sip:Alice@147.214.199.3>;tag=1220804646377
Max-Forwards: 70
Content-Length: 22
Cseq: 1 MESSAGE
Contact: sip:147.214.199.3:6061;transport=UDP
To: "SoccerChat"<sip:SoccerChat@147.214.199.3>
Content-Type: text/html
Via: SIP/2.0/UDP 147.214.199.3:6061;branch=z9hG4bK-1220804644969
Call-Id: 1220804646377
```

Who won the last game?

```
# Assigning user Alice to SipApplicationSession with key = SoccerChat
# ServletContext.log():Received message : Who won the last game? in
SAS with id = 10,10,SoccerChat/SipAppKeySample
```

```
MESSAGE sip:SoccerChat@147.214.199.3 SIP/2.0
From: "Chris"<sip:Chris@147.214.199.3>;tag=1220805161378
Max-Forwards: 70
Content-Length: 23
Cseq: 2 MESSAGE
Contact: sip:147.214.199.3:6061;transport=UDP
To: "SoccerChat"<sip:SoccerChat@147.214.199.3>
Content-Type: text/html
Via: SIP/2.0/UDP 147.214.199.3:6061;branch=z9hG4bK-1220805161376
Call-Id: 1220805161378
```

I think it was Arsenal!

```
# Assigning user Chris to SipApplicationSession with key = SoccerChat
# ServletContext.log():Received message : I think it was Arsenal! in
SAS with id = 10,10,SoccerChat/SipAppKeySample
```

However, a message in a totally different context would yield the following:

```
MESSAGE sip:Basketball@147.214.199.3 SIP/2.0
From: "Kristoffer"<sip:Kristoffer@147.214.199.3>;tag=1220805422864
Max-Forwards: 70
Content-Length: 40
Cseq: 3 MESSAGE
Contact: sip:147.214.199.3:6061;transport=UDP
To: "Basketball"<sip:Basketball@147.214.199.3>
Content-Type: text/html
Via: SIP/2.0/UDP 147.214.199.3:6061;branch=z9hG4bK-1220805422861
Call-Id: 1220805422864
```

What do you think about Kobe last night?

```
# Assigning user Kristoffer to SipApplicationSession with key = Basketball
# ServletContext.log():Received message : What do you think about
Cobe last night? in SAS with id = 10,10,Basketball/SipAppKeySample
```

The algorithm used by the annotation is defined by the developer but always produces a string based on the incoming SIP request. Note that it only needs to produce a string and does not have to consider the "SipServletRequest" given as input parameter. An implementation could very well return the weekday as a string resulting in different "SipApplicationSession" objects, one for every day in the week.

After receiving the SIP application name to which a new request is being dispatched by the Application Router, the container will search the application classes for the @SipApplicationKey annotation. If it is found, it invokes that method and supplies the SIP request as input. The output is always a string as a result of the developer-defined processing that has taken place. For example, the developer algorithm might take the request object and extract the Request-URI, looking for a specific value like "sip:chat_room@example.com." The string returned to the container would always be the same for that SIP URI. So the first time a request arrives with the Request-URI of "sip:chat_room@example.com," a new Application Session would be created to associate the SIP protocol session (SIP Session). Sometime later, another user sends a SIP request directed at the same SIP URI. The second time, the container would again look for the @SipApplicationKey annotation and provide the SIP request as input. The output would be a key in the form of the string that represents the Application Session that should be associated with the new second SIP request. The container then looks at its list of currently active Application Sessions. If the key created as a result of the "@SipApplicationKey" algorithm already exists as an active Application Session, then the container associates the new protocol session with it. If no existing active Application Session exists, the container creates a new one.

The power of this mechanism lies in the fact that the algorithm for generating the Application Session key lies in the hands of the developer and can be as simple or complicated as required. It shows how two unrelated SIP dialogs can be grouped together by the application key mechanism.

2.1.6.2 Session Targeted

The session-targeted mechanism occurs at a different stage of the application selection process and is intended to ensure the semantics of the special SIP headers Join and Replaces are met. SIP Join and Replaces are quite unusual in that they are SIP headers of an INVITE request that carry information relating to existing protocol sessions. In short:

- The SIP Join header is used to join a new SIP protocol session with an existing protocol session. The Join header appears in an INVITE request and contains the SIP protocol identifiers of the call it wishes to join. For example, it models the "barge-in" function where user A is in a SIP call with user B. User C sends a call to user A instructing that it also be included in this call, resulting in a three-way call among users A, B, and C. An example SIP INVITE message would look like the following (some parts of the SIP message have been left out):

```
INVITE sip:kristoffer@sipservlet_example.com SIP/2.0
To: <sip:kristoffer@sipservlet_example.com>
From: <sip:chris@sipservlet_example.com>
Call-Id:892374@sipservlet_example.com
CSeq: 1 INVITE
Contact: sip:chris@pc.sipservlet_example.com
Join: 789424@sipservlet_example.com;to-tag=xyz;from-
tag=abc
```

- The SIP Replaces works in the same way as Join except that, instead of adding a new SIP protocol session to an existing call, it replaces it. This is used for features such as call pickup. For example, user A is on a call with user B. User C is able to send an INVITE request containing the SIP protocol session identifiers to user A instructing that a new session be setup with user C and the CALL with user B be terminated (replaced). An example SIP INVITE message would look like the following (some parts of the SIP message have been left out):

```
INVITE sip:kristoffer@sipservlet_example.com SIP/2.0
To: <sip:kristoffer@sipservlet_example.com>
From: <sip:chris@sipservlet_example.com>;tag=89320u88
Call-Id:442374@sipservlet_example.com
CSeq: 1 INVITE
Contact: sip:chris@pc.sipservlet_example.com
```

```
Replaces: 832744@sipservlet_example.com;to-tag=xyz;
from-tag=abc
```

Looking at both Join and Replaces, it is obvious that for such requests to be successful an incoming request should be dispatched to an appropriate Application Session that is responsible for SIP protocol sessions that are referenced in both Join and Replaces SIP headers. Not providing a mechanism to associate such requests with an application and simply relying on it being part of normal procedures leaves success to chance and is almost certain to fail in some scenarios. It is for this reason that the session-targeted mechanism is used to ensure SIP INVITE requests that contain Join and Replaces SIP headers are dispatched correctly. It is not only the specific application that has to be targeted but also the correct "SipSession" object representing the actual SIP UA that is being invoked.

The session-targeted mechanism works at a slightly different level to the session key. Remember that the session key is applied only on a request once it has been sent by the container and has arrived at a SIP Servlet application. Only then is the SIP Protocol Session associated with an existing Application Session within the application deployed in the container. To ensure that INVITE requests containing SIP Join and Replaces headers are processed properly, the container has to ensure that the correct application is selected in the first place, and so it occurs much earlier in the application selection procedures.

On receiving an INVITE request, the container will inspect Join and Replaces SIP headers and determine using the SIP protocol identifiers (from the Join/Replaces header) whether it exists and which application it is associated with. We briefly mentioned previously that the container recursively interacts with a logical entity called an Application Router to determine which SIP Servlet archive should be visited. The container carries out the following additional steps when interacting with the Application Router to make sure the appropriate application is invoked to service the Join and Replaces headers: The container calls the Application Router API interface requesting the application that should receive the new INVITE request. The container notices that the new INVITE request contains either a Join or Replaces SIP header. On making the request to the Application Router, the container will include the information from the Join or Replaces header in a "SipTargetedRequestInfo" object along with the application name it located when the SIP INVITE arrived. As a result, the Application router has the appropriate contextual information to return the application name that is responsible for the Join or Replaces header. The container is then able to associate the new SIP INVITE request with the appropriate Application Session, and the application logic is free to carry out Join and replaces style operations.

You may be asking yourself why, on receiving the INVITE request, the container went to the trouble of looking up the appropriate Application Session and then provided that information to the Application Router, only to be returned the same application name. Why not just do the initial lookup and associate the new INVITE that contains the Join or Replaces header without consulting the Application Router? The Application Router is a vital concept as part of SIP Servlet 1.1 architecture and is the primary decision maker on which application should be invoked. To bypass the Application Router would not be recommended and is not in keeping with the application composition model. After all, it might be that invoking a security or monitoring application is required before the actual target of the Join and Replaces header application. To bypass the Application Router would result in circumventing such applications. The container using the Application Router interface to supply appropriate contextual information leaves the ultimate decision with the appropriate controlling entity (the Application Router). This maintains a consistent application selection process regardless of individual semantics that produce new selection requirements. Circumventing applications for targeted SIP requests was the very reason that the previously described Encode URI mechanism was depreciated from the specification.

One note worth considering is that the container behaves in exactly the same way handling both Join and Replaces. It is up to the Servlet programmer to terminate the replaced SIP Session and generate a properly formatted SIP BYE message. This is due to the fact that an application might want to include a header or body into the massage and that a SIP BYE might yield in an authentication response that needs to be handled in application space.

Having three potential mechanisms is obviously not ideal, so it's important that a container implement a clear set of rules in case a conflict occurs. SIP Servlet 1.1 specifies a priority for the three mechanisms that enables a container to manage such conflicts. On receiving a request, the container identifies the request with a Join/Replaces header (and it supports this optional mechanism), and it is able to locate the associated Application Session—the container uses this information to consult the Application Router even if an Encode URI was present in the SIP request. Progressing a step further, if a request is dispatched to an application using the Join/Replaces targeting mechanism, the container should not use the @SipApplicationKey annotation method for associating the protocol session with an existing Application Session. It should associate the Join/Replaces header with the appropriate Application Session, the one that was identified even before the Application Router was consulted. If a request is dispatched to an application due to the presence of an Encode URI, the container should still use the @SIP ApplicationKey annotation present to associate the SIP protocol session with the Application Session. Otherwise, use the Encode URI present in the SIP message to associate the SIP protocol session with the Application Session.

2.1.7 Session Utilities

The importance of integrating SIP Servlet technology into larger applications and making it part of wider architectures such as JEE results in a number of interesting facilities that a container must support. The injection, using Java annotations, of SIP Session Utilities, a Timer Service, and an instance of a SIP Factory are the three primary enablers for wider application integration. This section will cover the injection of SIP Session Utilities, while the next sections will cover SIP Factory and Timer Service. It should be noted that a detailed section on SIP Servlet-specific Java annotations is included later in the book.

SIP Session Utilities (or the "SipSessionsUtils" Java interface) provides a mechanism for applications to look up and use existing SIP Application Sessions. The lookup may be carried out by the SIP Servlet application itself, which would result in a local operation, or it might be by a third-party application not located on the SIP Servlet container, for example, an EJB hosted on a separate EJB container. The transparency of Java annotations means that the application developer using "SipSessionsUtils" cannot tell by the code whether the injection is local or remote. If it is remote, an appropriate naming directory function such as the Java Naming and Directory Interface (JNDI), which is part of the JEE architecture, can used. SIP Application Sessions are looked up based on a unique identifier that is generated by container.

Let's step through an example to illustrate how and why an application might want to look up and use a SIP Servlet Application Session. Let's imagine we have a fictional JEE application that has an EJB part to carry out the majority of business logic, which is hosted in an EJB container separate to the SIP Servlet container. At some point during the operation, the EJB has been instructed to add the boss to a conference call at a certain time. The conference call commences on the SIP Servlet container with lots of participants dialing in and discussing many topics. As a result, the SIP Servlet container will have an active Application Session representing the conference call and any number of SIP protocol sessions (SIP Session) representing each user. For every Application Session created by a SIP Servlet container, an associated, unique identifier is also created that maps 1:1 and lives for the duration of the Application Session. The unique identifier can be obtained by applications that call the "getId" method on the "SipApplication Session" interface of the SIP Servlet API. Time lapses and the EJB instance obtain the unique Application Session identifier by injecting an instance of the "Sip SessionsUtils." This is achieved using the @Resource annotation, which is defined in the Common Annotations for Java Platform (JSR 250) [12]. The Java code in the EJB would look something like this:

```
@Resource
SipSessionsUtil util_instance;
```

An instance of the "SipSessionsUtils" has now been passed locally into the local instance variable named "util_instance" (note that the specific application instance would have been specified using the JNDI entry used for the injection "sip/<appname>/SipSessionsUtil"). Now that our EJB has a local handle on the

"SipSessionsUtils" associated with the SIP Servlet container, it can look up the Application Session based on the previously discussed unique identifier. How the EJB obtains the unique identifier is out of the scope of this discussion—it could, for example, be as a result of a database lookup. The EJB can now use the unique identifier to get access to the Application Session by calling the "getApplicationSessionById" method on the local instance of the "SipSessionsUtils" interface (which we previously injected) and passing in the unique identifier. This would look something like this:

```
newAppSession = util_instance.getApplicationSessionByID(832748sad);
```

At this point, an instance of the Application Session associated with the unique identifier has been assigned to the local "newAppSession" instance. The EJB application now has control of the Application Session and can use the full range of facilities available as part of the SIP Servlet API to generate the appropriate SIP protocol signaling to add the boss to the conference call (the main elements of the SIP Servlet API will be discussed later in the book).

As we touched on earlier in this section, the ability to inject an instance of "SipSessionsUtil" into an application is transparent to the developer and can occur either locally, within a SIP Servlet class, or remotely, as in our example. It should also be noted that a local SIP Servlet class can also obtain an instance of "Sip SessionsUtil" using the Servlet Context. The Servlet Context is a concept that is inherited from the base Servlet specification and provides "a servlet's view of the SIP application within which the servlet is running" [13]. It is used for holding configuration values and essential static application data and is discussed in more detail later. Injecting an instance of "SipSessionsUtil" from the Servlet Context would look like this:

```
SipSessionsUtil util_instance = (SipSessionsUtil)getServletCon-
text().getAttribute("javax.servlet.sip.SipSessionsUtil");
```

An instance of the "SipSessionsUtil" is injected locally from the context into "util_instance." Because this is acting locally and is within the application, there is no need to specify the application name (as we did using Java Naming and Directory Interface (JNDI) in the example); it's a totally local operation. This would have been no use in our example, as the "SipSessionsUtil" was injected into an EJB sitting in a remote EJB container.

While injecting "SipSessionsUtil" is a very powerful tool, a container has to be wary of potential conflicts between remote and local actions being taken on the same Application Session. So, in our previous example, while the remote EJB was attempting to directly act on the SIP Application Session, the local application may also have been attempting operations. There is no formal model for multiple request threads acting on SIP Servlet application session objects, so the container should ensure that synchronized access is maintained to ensure consistency. It is hoped that in future versions of the SIP Servlet architecture a common asynchronous mechanism will be introduced.

Note One nice development pattern that the "SipSessionsUtil" provides is that a SIP application could use a "SipApplicationSession" as a shared storage. Initialization of a new "SipApplicationSession" could be created with the "SipFactory" method "create ApplicationSessionByKey." Then, anywhere in the application, a "SipSessionsUtil" can be annotated and a reference to the "SipApplicationSession" can be found based on the key it was created with.

2.1.8 SIP Factory

The SIP Factory (known as "SipFactory" Java interface) is the second of the utilities discussed in this section, and its name very much describes its function. It is an interface that is used by an application to create new instances of certain important interfaces that exist within the core SIP Servlet API. In other words, it acts as a factory for producing all the useful artefacts that application developers use in the SIP Servlet API on a regular basis. The following provides a general list of methods that an application can access from obtaining an instance of the SIP Factory of the application:

- *createAddress*—an "Address" object in the SIP Servlet API is a representation of SIP address that is found in a number of key SIP protocol headers such as "To" and "From." It is a convenience object for SIP application developers, because it is a container for not only the SIP URI that appears in the SIP headers but also display name, if present, and any parameters. This method allows for "Address" objects to be created when the SIP Factory is supplied with appropriate parameters in the method call (e.g., the SIP URI and display name). The address interface basically represents the "name-addr" format specified in SIP 2.0 RFC 3261 [2].

- *createApplicationSession*—We have already talked about SIP Application Sessions and how they are created on receiving new SIP protocol messages such as INVITE. This method allows the SIP Factory to create a brand-new instance of an application without the need for any SIP protocol signaling. This might be used, for example, to create a conference that participants dial in to and that needs to be created before the first one enters.

- *createApplicationSessionByKey*—This method is identical to the previous "createApplicationSession" method, with the exception that the Application Session is created using a specific key. You might remember that we discussed the session key mechanism for targeting SIP Application Sessions using the @SipApplicationKey Java annotation. Using this method ensures that a new SIP Application Session is created, which allows incoming new requests to easily be associated with an existing SIP Application Session.

- *createAuthInfo*—It was recognized that programmatically creating appropriate authentication interactions, such as those defined for SIP Digest in RFC 3261 [2], using SIP is quite a difficult task. This method creates a configurable object (called "AuthInfo") that can be used in the SIP Servlet API for easily attaching appropriate authentication information to appropriate SIP messages.

- *createParameterable*—The "Parameterable" interface is another convenience object within the SIP Servlet API that allows for ease of manipulation of SIP header values. The "createParameterable" method call creates a new "Parameterable" object based on the input string provided.

- *createRequest*—Once you have an active SIP Application Session (either through the "createApplicationSession" method defined for this SIP Factory or as a result of new SIP protocol signaling), a developer is able to create new SIP protocol requests using this method (e.g., such as a SIP INVITE). It should be noted that using the SIP Factory in this way creates totally independent SIP protocol signaling interactions that are unrelated to previous SIP Sessions. The ability to send new requests within an existing SIP Session is discussed later in the book.

- *createSipURI*—The method for inputting the names of the appropriate user and host part to create a new SIP URI. For example, including a user part of "chris" and a host part of "sipservlet_example.com" would return a SIP URI object of "sip:chris@sipservlet_example.com."

- *createURI*—Similar to previously method call except a string value is passed as a parameter and a general URI object is returned, as defined in RFC 2396 [14]. This object can equally be used in SIP requests and "Address" objects but provides a more generalized format.

All of these are well-known constructs to SIP Servlet application developers, and the SIP Factory interface provides an easy and consolidated convenience function for their creation. These constructs will become more familiar later in the book, when the focus will switch from the container to application creation and the SIP Servlet API.

As a utility, the SIP Factory is used within applications in a manner similar to the previously defined "SipSessionsUtil." It can be made available locally within a SIP Servlet class using a Servlet Context (which we introduced in the

"SipSessionsUtil" section) attribute called "javax.servlet.sip.SipFactory." It would look something like this:

```
SipFactory local_instance = (SipFactory)getServletContext().
getAttribute(SIP_FACTORY);
```

An application can then use the object "local_instance," taken from our example, to call any of the methods defined previously.

An instance of the SIP Factory can also be injected using the Java-based @Resource annotation that was also used for "SipSessionsUtil." This can be called both locally in the SIP Servlet class and externally from a larger JEE application. It would look something like

```
@Resource
Private SipFactory local_instance
```

where an instance of SIP Factory is injected into the local object "local_instance."

If we progress our previously discussed example forward from the Session Utilities section of the book, we are attempting to add the boss to an existing conference call at a certain time. The time has elapsed, and we now want to call the boss for inclusion in the conference call with colleagues. We previously used the "SipSessionsUtil" by injecting an instance into our EJB-based application, which allowed us to retrieve the appropriate Application Session based on a unique identifier. We are now able to use the previously obtained Application Session with the SIP Factory to achieve the required SIP protocol signaling to add the boss the conference. For example, using the previously described SIP Factory methods "createAddress" and "createSipURI," we can specify the appropriate SIP URIs to contact the boss. We can then use the previously obtained Application Session in conjunction with the SIP Factory to create a new request to the boss using the "createRequest" method. We have now contacted the boss using the previously obtained Application Session and the SIP Factory.

2.1.9 Timer Service

The Timer Service is the third and final container utility covered in this section and provides a mechanism for applications to "schedule timers and receive notifications when timers expire" [13]. Creation of specific timers is achieved using the Timer Service (TimerService Java interface) while the Timer Listener (TimerListener Java interface) is a callback interface that can be implanted by applications and is invoked when a previously created timer fires. A Servlet Timer object (ServletTimer) is used by the Timer Service to pass appropriate information to the Timer Listener.

The Timer Service only has two variations on the "createTimer" method, which is used to create and instantiate an application timer. The first instance of "createTimer" has the following parameters, which are passed in as configuration:

- *appSession*—Specifies the Application Session that the new timer should correlate with.
- *Delay*—The delay in milliseconds before the timer will fire.
- *isPersistent*—A Boolean value that indicates if the timer should continue following a shutdown of the system.
- *info*—A serializable object that enables applications to store generic application data that is returned when the timer fires.

The second "createTimer" method in the Timer Service interface consists of all the previously described methods and the following additions:

- *fixedDelay*—A Boolean parameter that allows both fixed rate and fixed delay mode.
- *Period*—The period in milliseconds between expiration of timers.

In similar way to both the Session Utilities and SIP Factory interfaces, the Timer Service can be injected locally in the SIP Servlet class using a Servlet Context attribute, which would look like the following:

```
TimerService local_instance = (TimerService)getServletContext().
getAttribute(TIMER_SERVICE);
```

Or it can be injected either locally in the SIP Servlet class or as part of a larger application using the Java @Resource annotation, which would look like this:

```
@Resource
TimerService local_instance;
```

An application that wishes to be notified when a configured timer expires has to implement the SIP Servlet API Timer Listener (TimerListener Java interface). This callback interface is triggered when a timer expires and the single method "timeout" is called by the container. The "timeout" method returns an object of type "ServletTimer" that contains all of the relevant contextual information associated with the timer. The "ServletTimer" object has the following methods that an application can use on receiving a call to the "timeout" method on the Timer Listener interface:

- *cancel*—Cancel the active timer.
- *getApplicationSession*—Provides the SIP Application Session associated with the timer.
- *getId*—Returns the unique identifier that has been assigned to this specific timer.
- *getInfo*—The ability to retrieve the serialized object of application-specific information that was inserted during the instantiation of the timer.
- *getTimeRemaining*—The amount of timer, in milliseconds, that remains until the timer expires.
- *scheduledExecutionTime*—Provides the expiration time of the timer.

The Timer Listener can be used by applications in one of two ways. It can make use of the @SipListener Java annotation for receiving the timer events, or it can use the deployment descriptor <listener> element, which has a child element <listener-class>, which is hard coded to inform the container where the listener class exists.

2.2 Container Convergence

A theme running through these first chapters has been the origins of the SIP Servlet architecture in conjunction with the base Servlet specification. This parallels the similarities drawn between the origins of the SIP signaling protocol in the HTTP protocol. It is natural to see why the SIP Servlet architecture has been closely aligned with the HTTP Servlet architecture, which provides a convenience mechanism for those classes of applications that require both technologies (e.g., a simple click-to-dial application that has both a Web and SIP function). We have also discussed that SIP Servlet technologies' ultimate goal is to become part of the core JEE architecture. Mechanisms and examples have already been introduced that signify a move toward making integration of larger JEE-based applications easy. Remember the example in which we used an EJB to include the boss in a preexisting conference call—it is a classic example of SIP Servlet technology's being integrated as part of a larger JEE application.

It is for these reasons that convergence of SIP Servlet technology with other technologies is considered extremely important. Yes, it is a valid architecture to deploy a SIP Servlet container as a stand-alone entity, but the real power of the technology is apparent when used with complementary technologies. The term *convergence* plays a major role in the SIP Servlet architecture and is defined by two main categories:

- *SIP Servlet container and HTTP Servlet container convergence*—Represents a tight integration of both a SIP Servlet container and an HTTP container with a shared Servlet Context. This allows an application both to contain SIP Servlet and HTTP Servlet deployment descriptors within the same deployment unit and also to have direct interactions with each other. It should be noted that tightly coupled convergence with an HTTP Servlet container is not required by an implementation of the SIP Servlet architecture. The SIP/HTTP convergence mechanisms defined in the SIP Servlet specifications are used for convenience as they cover a large majority of simple Web/telecom applications (e.g., click-to-dial).

- *SIP Servlet container and JEE convergence*—More loosely coupled than the previous convergence with HTTP, it refers to a SIP Servlet's being part of a larger application that can include various other well-known JEE functions, such as EJB and Web services. The example previously discussed involving an EJB's triggering a SIP protocol interaction to include a boss in a conference call, is an example of such convergence. SIP Servlet 1.1 has evolved to provide more convenience to such convergence with the introduction of Java-based annotations in conjunction with Session Utilities, SIP Factory, and Timer Service, which were all previously covered in this book.

It's important when dealing with convergence of SIP Servlet technology that it be clear which of the previous two definitions is appropriate. Just using the term *convergence* does not supply enough context and can cause confusion.

The remaining two sections will take a closer look at the two types of SIP Servlet container convergence.

2.2.1 HTTP Container Convergence

The tight convergence with an HTTP Servlet container is reasonably well understood, and a number of convenience functions are provided by a SIP Servlet container that remove any potential complexity. Having a common mechanism for such interactions also improves chances of interoperability and encourages further innovation and deployments.

Creating applications that potentially have a number of components with varying technologies results in the need for a set of packaging guidelines. The following lists the associated deployment units that are used for the technologies in question:

- *.sar file extension*—Defined by the SIP Servlet architecture, it is the package structure for a SIP Servlet application.

- *.war file extension*—Defined by the HTTP Servlet architecture, it is the package structure for a HTTP Servlet application.

- *.ear file extension*—Defined by the JEE architecture, it is the package structure for a JEE application. It should be noted that a JEE ".ear" application archive is a higher level container for all JEE technologies and so can contain both HTTP Servlet and SIP Servlet parts.

A converged application in this particular context can be packaged in either a ".sar" or ".war" file or can be packaged as part of a larger, JEE-based application when included in an ".ear" file.

The basis for a SIP Servlet application to be tightly converged is based on the sharing of the Servlet Context. We mentioned previously that the Servlet context provides a view of the application and contains static metadata relating to the application (see later in the book for more detail on Servlet context data). It is important that both SIP Servlets and HTTP Servlets have a consistent view of the application. Figure 2.11 is a high-level representation of a shared context between both HTTP and SIP Servlets.

The ability to use functions such as Session Utilities (for obtaining SIP Application Sessions), SIP Factory (for creating SIP Signaling) and Session Timer (for creating timer tasks) provided by the SIP Servlet container have already been discussed. In the case of a shared Servlet context, such functions can be used as if being called from within the SIP Servlet class. This can be achieved using either the Servlet Context attributes for each function or the appropriate Java @Resource annotation. So, for example, you would have an HTTP Servlet class that is able to directly call the context attribute to obtain a Session Utilities instance (or use Java annotation). Even though it is outside of the SIP Servlet class, it is considered a local call due to the single application archive (either .sar, .war, or .ear) and shared Servlet Context. While entirely legal, it is not the recommended mechanism for obtaining a SIP Application Session from within an HTTP Servlet Class. The SIP Servlet API offers an interface called "ConvergedHTTPSession," which has a "getApplication Session" method for obtaining a SIP Application Session. A shared SIP and HTTP Servlet container makes the "ConvergedHTTPSession" available to HTTP Servlet-based applications, which can then use the "getApplicationSession" method to

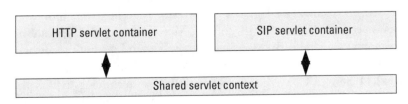

Figure 2.11 SIP and HTTP shared context.

obtain the associated SIP Application Session. If no SIP Application Session exists, then one is immediately created and returned to the HTTP Servlet application for further processing.

A converged Web container makes sure that the "HttpSession" class representing a cookie session is also implementing the "javax.servlet.sip.Converged-HttpSession" interface. In SIP Servlet 1.0, this was not specified, so how a container connected the "HttpSession" to a "SipApplicationSession" was implementation-specific. Generally, if an "HttpServletRequest" has a session and initiated a new SIP request with the SipFactory, the container made the association. Now, in SIP Servlets 1.1, by safe casting the instance of SIP Factory and then by calling "get ApplicationSession," this gives the application developer full control.

Using a simple click-to-dial application as an example, the following steps would occur:

1. A simple HTTP Servlet application has a Web page with an input field to identify the user that is to be called. The user enters a valid SIP URI and clicks the appropriate button to proceed.

2. The HTTP Servlet application uses the "ConvergedHttpSession" that has been made available by the container and invokes the "getApplication Session" method to obtain the related SIP Application Session.

3. The HTTP Servlet application is now able to obtain the appropriate SIP Factory for the application and create the SIP protocol-level signaling to connect the users together using the "createRequest" method from the SIP Factory interface.

2.2.2   JEE Container Convergence

While convergence with an HTTP Servlet container provides a nice convenience for application developers who want to write simple applications, the real power and flexibility of SIP Servlet technology is evident when discussing convergence with JEE. A clear division of responsibilities between the telecom-based infrastructure and the enterprise business world provides a highly scalable architecture.

The division of responsibility, as shown in Figure 2.12, allows specific specialist technologies to handle appropriate application logic relevant to the area of expertise. As illustrated by Figure 2.12, the JEE is dealing with business technologies such as Web services, EJB, and Java Message Service (JMS), while the SIP Servlet container handles SIP signaling logic and lower level multimedia session-related logic. This is not saying that this book is promoting convergence with JEE over HTTP Servlet. Convergence with both types of technology has its place in the SIP Servlet architecture and represents a totally different set of deployments. It must also be remembered that a SIP Servlet container does not have to converge with either of the two technologies discussed and can act as a stand-alone entity

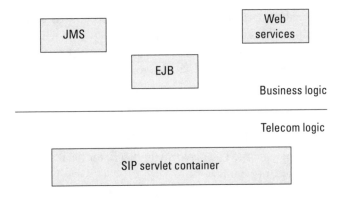

Figure 2.12 Division of responsibility.

providing just SIP-based multimedia logic. It is fair to say, however, that convergence with JEE can cover the entire range of deployments of a converged HTTP container and more. It is not fair to say the reverse, that a converged HTTP container has the same range of flexibility. On the flip side, it is overkill to require a full-blown JEE Application server to create a very simple click-to-dial application, and so a converged HTTP Servlet container is perfect.

The majority of details surrounding convergence with JEE have been discussed directly and indirectly already and will be continue to be discussed as a general theme. In fact, not only has the book looked in detail at SIP Servlet container functions such as remotely using Session Utilities, SIP Factory, and Timer Service, it has even given a simple example of an EJB's triggering SIP Servlet logic (recall the example of adding the boss to a conference call at a certain time).

As specified by JEE, a converged SIP Servlet application will be bundled in an enterprise archive file format (an ".ear" file, e.g., MyConvergedApp.ear). For more information on how ".ear" files are structured and packaged, the reader should take a look at the latest version of the core JEE specification. The Enterprise Archive format acts as boundary for converged application. This includes the availability of previously mentioned facilities such as Session Utilities, SIP Factory, and Session Timer. That is not to say that a stand-alone container not part of a JEE application server cannot offer these functions to third-party servers; however, in the context of a JEE Application Server, the Enterprise Archive acts as the boundary for access.

2.3 Security

Whether a container's role is one of the three we have previously discussed (JEE converged, HTTP Servlet converged, or stand-alone), security plays a major role. A SIP Servlet container needs to be able to provide a common security architecture that allows application developers the ability to clearly state security constraints associated with the application. Those then responsible for actually deploying and managing a live application in the network need to be able to ensure that these security constraints are being effectively enforced. Add to that the requirement that an application could be sold and deployed to a wide range of customers and deployed on a wide range of SIP Servlet containers produced by many vendors. It is for this reason that the SIP Servlet architecture has a common security framework that is implemented and enforced by a container, which should ensure consistency across a wide range of deployments. The deployment descriptor file is the primary vehicle for specifying the security framework.

The security requirements for a SIP Servlet container are quite wide ranging. While, on one hand, a container could be deployed in a totally closed network where access is controlled at the perimeter, the security requirements will not be nearly as strong as if it is deployed in an open environment as part of a JEE application server deployment. On a multitenant system, there could be a plethora of potential users attempting to get access to container resources. It is for this reason that the SIP Servlet architecture splits security properties into a number of core requirements, which are specified in the SIP Servlet specification [13] and listed here:

Authentication: The means by which communicating entities prove to one another that they are acting on behalf of specific identities that are authorized for access [13].

Access control for resources: The means by which interactions with resources are limited to collections of users or programs for the purpose of enforcing integrity, confidentiality, or availability constraints [13].

Data integrity: The means used to prove that information has not been modified by a third party while in transit [13].

Confidentiality or data privacy: The means used to ensure that the information is made available only to users who are authorized to access it.

The security architecture spans both elements that will appear in the deployment descriptor (described as "declarative security" in the SIP Servlet architecture) and can appear in the actual application code using specific methods (described as "programmatic security" in the SIP Servlet architecture). It is a combination of both these mechanisms that allows for appropriate security constraints to be specified by an application and enforced by a container.

One area in which the security mechanism needs to be aligned is the case where a SipServlet requires authentication from the client executing the code. This is done by issuing a 401 SIP response asking for valid credentials. The SIP client making the initial call then provides the right data, and the Servlet can validate its authenticity. So far, this is a normal description of how a SIP entity would behave. The important part is that the JEE application server should behave as one entity. Any call generated from the SIP Servlet has to pass on the authentication and authorization information collected by the SIP container. If a call is made to EJB, then there should not be a need for a new authentication for accessing a specific resource. Only the authorization step is run, since we already know the identity of the SIP client on whose behalf it runs.

To protect a Servlet, the sip.xml configuration file would look like the following:

```xml
<?xml version="1.0" encoding="UTF-8"?>
<sip-app xmlns:xsi="http://www.w3.org/2001/XMLSchema-instance"
xsi:schemaLocation=http://www.jcp.org/xml/ns/sipservlet
http://www.jcp.org/xml/ns/sipservlet/sip-app_1_1.xsd" version="1.1">
        <app-name>UASAuthSample</app-name>
        <servlet>
          <servlet-name>UASAuthServlet</servlet-name>
          <servlet-class>net.sipservlet.sample.AuthServlet</servlet-class>
        </servlet>
        <servlet-mapping>
          <servlet-name>UASAuthServlet</servlet-name>
          <pattern>
            <equal>
              <var>request.method</var>
              <value>REGISTER</value>
            </equal>
          </pattern>
        </servlet-mapping>
        <security-constraint>
          <resource-collection>
            <resource-name>UASAuthServletProtector</resource-name>
            <servlet-name>UASAuthServlet</servlet-name>
            <sip-method>REGISTER</sip-method>
          </resource-collection>
          <auth-constraint>
           <role-name>authenticated</role-name>
          </auth-constraint>
        </security-constraint>

        <login-config>
          <auth-method>DIGEST</auth-method>
          <realm-name>sipservlet.net</realm-name>
        </login-config>
</sip-app>
```

Here the "UASAuthServlet" is protected for any calls to the REGISTER SIP method. It will do a digest 401 authentication. If the authentication step is done, it will ensure that the authenticated user has the "authenticated" role assigned to it.

If not, then a 403 SIP response would get generated. If both the authentication and authorization are successful, then the actual Servlet would get invoked. Here is how the Servlet code would look:

```
public class AuthServlet extends SipServlet {
  @Override
  protected void doRegister(SipServletRequest req) throws Servlet
Exception, IOException {
    log(req.getRemoteUser());
    log(req.getUserPrincipal().toString());
    log("Role = "+req.isUserInRole("authenticated"));
    SipServletResponse resp = req.createResponse(200);
    resp.send();
  }
}
```

The Servlet prints the authentication information and simply creates a SIP 200 response for the registration. Running it would yield the following printouts:

```
[#||INFO|sun-comms-
appserver1.0|javax.enterprise.system.container.web|_ThreadID=23;_
ThreadName=SipContainer-serversWorkerThread-5060-7;|PWC1412:
ConvergedContextImpl[/AuthServer] ServletContext.log():stoffe|#]

[#||INFO|sun-comms-
appserver1.0|javax.enterprise.system.container.web|_ThreadID=23;_
ThreadName=SipContainer-serversWorkerThread-5060-7;|PWC1412:
ConvergedContextImpl[/AuthServer] ServletContext.log():stoffe|#]

[#||INFO|sun-comms-
appserver1.0|javax.enterprise.system.container.web|_ThreadID=23;_
ThreadName=SipContainer-serversWorkerThread-5060-7;|PWC1412:
ConvergedContextImpl[/AuthServer] ServletContext.log():Role = true|#]
```

Note   Each application server has its own way of defining where the user name and password information are collected. Normally, it is done using some version of JAAS (Java Authentication Authorization System). Running this on SailFin requires one table with the user name and password and also another table with the user-to-group mapping. To make it easier in deployment, it also needs a "sun-sip.xml" descriptor file to map between the group and the roles. This is due to the fact that, while one person creates the application, it can be another person who deploys it. As you see in the sample, checks for "isUserInRole" can be made programmatically without the mapping, but one might have to recompile the SIP application, which is not an acceptable approach [15].

This was security mechanism acting as a User Agent Service (UAS), but there is a corresponding pattern for writing SIP User Agent Client (UAC) code.

```
@javax.servlet.sip.annotation.SipServlet(name = "authServlet",
loadOnStartup = 2)
public class AuthSipServlet extends javax.servlet.sip.SipServlet {

  @Resource
  SipFactory sf;
  @Resource
  TimerService ts;

  private static final long serialVersionUID = 3978425801979081269L;
  //Reference to context - The ctx Map is used as a central storage
for this app
  ServletContext ctx = null;

  @Override
  public void init(ServletConfig config) throws ServletException {
    super.init(config);
    ctx = config.getServletContext();
  }

  @Override
  protected void doResponse(SipServletResponse resp) throws
javax.servlet.ServletException, java.io.IOException {
    SipUser user = (SipUser) ctx.getAttribute("sipuser");
    if (resp.getStatus() == SipServletResponse.SC_UNAUTHORIZED ||
       resp.getStatus() == SipServletResponse.SC_PROXY_AUTHENTICATION_
REQUIRED) {
      AuthInfo info = sf.createAuthInfo();
      info.addAuthInfo(resp.getStatus(), user.getRealm(),
user.getUser(), user.getPassword());
      SipServletRequest req = resp.getSession().createRequest
(resp.getMethod());
      req.addAuthHeader(resp, info);
      req.pushRoute(sf.createAddress(user.getOutboundproxy()));
      req.getSession().setHandler("authServlet");
      if (resp.getRequest().getContent() != null) {
        req.setContent(resp.getRequest().getContent(), resp.get
Request().getContentType());
      }
      if (resp.getRequest().getExpires() > 0) {
        req.setExpires(resp.getExpires());
      }
      req.send();
    } else if (resp.getStatus() == SipServletResponse.SC_OK) {
      log("Auth OK : " + resp.getHeader("Contact"));
      if ("REGISTER".equals(resp.getMethod())) {
        String serviceRoute = resp.getHeader("Service-Route");
        if (serviceRoute != null && serviceRoute.length() > 0) {
          user.setServiceRoute(serviceRoute);
```

```
                    }
                }
            }
        }
}
```

The user information is kept in a JavaBean "SipUser" where username, password, and realm are kept. This Servlet would normally be set to deal with responses. If a SIP server generates a 401 or 407 authentication request, then in "doResponse" method this Servlet would create the right authentication information. First an "AuthInfo" object is created with the help of the SIP Factory. Then the "AuthInfo" object is populated with the response code, username, password, and realm in the "addAuthInfo" method call. The last step is to append the "AuthInfo" to the request and send it back to the server. If the server accepts, then we would get a 200 response back. If any of the credentials are wrong or we do not have the right authorization, then we would receive a 403 response.

Note   that an entirely new request is created based on the "SipSession" of the response. This means that any extra SIP headers or SIP body from the original request needs to be copied from the original request to the one with "AuthInfo." We cannot add the "AuthInfo" to the initial because the "CSeq" would need to be increased. We cannot send the original request one more time.

References

[1]   SIP Servlet Specification, Version 1.1, JSR 289, Java Community Process, August 2008.

[2]   Rosenberg, J., et al., "SIP: Session Initiation Protocol," RFC 3261, Internet Engineering Task Force, June 2002.

[3]   Rosenberg, J., and H. Schulzrinne, "Reliability of Provisional Responses," RFC 3262, Internet Engineering Task Force, June 2002.

[4]   Roach, A. B., "Session Initiation Protocol (SIP)—Specific Event Notifications," RFC 3265, Internet Engineering Task Force, June 2002.

[5]   Campbell, B., et al., "Session Initiation Protocol (SIP) Extension for Instant Messaging," RFC 3428, Internet Engineering Task Force, December 2002.

[6]   Donovan, S., "The SIP INFO Method," RFC 2976, Internet Engineering Task Force, October 2000.

[7]   Rosenberg, J., "The Session Initiation Protocol (SIP) UPDATE Method," RFC 3311, Internet Engineering Task Force, September 2002.

[8]   Sparks, R., "The Session Initiation Protocol (SIP) Refer Method," RFC 3515, Internet Engineering Task Force, April 2003.

[9]   Niemi, A., "Session Initiation Protocol (SIP) Extension for Event State Publication." RFC 3903, Internet Engineering Task Force, October 2004.

[10] Mahy, R., and D. Petrie, "The Session Initiation Protocol (SIP) 'Join' Header," RFC 3911, Internet Engineering Task Force, October 2004.

[11] Mahy, R., B. Biggs, and R. Dean, "The Session Initiation Protocol (SIP) 'Replaces' Header," RFC 3891, Internet Engineering Task Force, September 2004.

[12] Common Annotations for the Java Platform, JSR 250, Java Community Process, May 2006.

[13] SIP Servlet Specification, Version 1.1, JSR 289, Java Community Process, August 2008.

[14] Berners-Lee, T., R. Fielding, and L. Masinter, "Uniform Resource Identifiers (URI): Generic Syntax," RFC 2396, Internet Engineering Task Force, August 1998.

[15] Binod's blog on SailFin Authentication: http://weblogs.java.net/blog/binod/archive/2008/09/md5_authenticat.html.

3

The SIP Servlet Application

This book has taken a good look at the roles and responsibilities of a SIP Servlet container, and it's now time to closely examine the role of an application that would utilize such container functionality. This section will focus on the main constructs that are used to both build and represent an application. The following section will focus more on the SIP Servlet API that is exposed to application developers when creating SIP Servlet applications. The two sections are closely related and will reference each other to aid in greater understanding.

3.1 SIP Servlet Packaging

This book has already touched on the packaging of SIP Servlet. A SIP Servlet application is wrapped up in a standard deployment unit that aligns with the JEE architecture. The deployment unit specified for a SIP Servlet application is defined as a Servlet archive and makes use of the ".sar" file extension. The ".sar" file extension has a standardized directory structure that allows containers to easily deconstruct and extract appropriate resources when deploying an application. The standard directory structure allows for interoperability when deploying applications on multiple vendor instances of a SIP Servlet container.

A directory folder appears in the root of a SIP Servlet application named "WEB-INF." This directory acts as container for everything that is related to an application. When a SIP Servlet container actively deploys an application, it will look straightaway for an existence of the "WEB-INF" directory so it can begin processing. The "WEB-INF" directory will contain a file called "sip.xml" that represents the deployment descriptor element of the application. (The details of the deployment descriptor can be found in Section 3.1.1 The "WEB-INF" directory then has a further number of subdirectories that contain the rest of the application resources. A "classes" directory contains the related SIP Servlet classes that are to be included by the applications class loader and made available at run time to the container. These would contain the relevant application logic that has utilized the SIP Servlet API. Finally, a directory called "lib" would also exist, which is used to include various Java archive files (.jar files) that are used by the application during run time.

Figure 3.1 illustrates the directory structure that is expected within a SIP Servlet application. It is worth noting that in this chapter the directory structure has only really been discussed in relation to the SIP Servlet archive format (.sar). Previously in the book we have introduced the concept that SIP Servlet applications can be converged and deployed with HTTP Servlet applications (using a web archive) and JEE applications (using an Enterprise Archive). The directory structure when included in either a Web archive (.war) or an Enterprise Archive (.ear) remains the same under the umbrella of the base "WEB-INF" directory. The structure of a SIP Servlet application mirrors that of an HTTP Servlet application, which also has a base "WEB-INF" directory and a deployment descriptor file called "web.xml" (as opposed to "sip.xml" in a SIP Servlet application). In the case where a SIP Servlet application is converged with an HTTP Servlet application, only a single "WEB-INF" directory exists, which is then shared by both the HTTP and SIP Servlet applications. Figure 3.2 shows an example of the single use of "WEB-INF."

The existence of a "sip.xml" and a "web.xml" signifies that this application archive has both SIP and HTTP Servlet parts. Due to the common root directory structure, the actual archive file extension type used by the converged application does not matter. For example, a SIP Servlet archive (".sar" file) can contain a

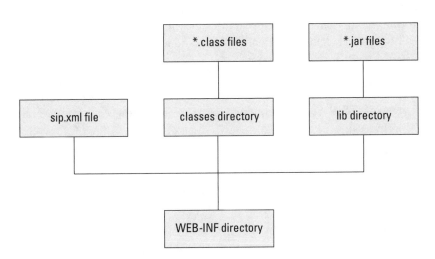

Figure 3.1 SIP Servlet directory structure.

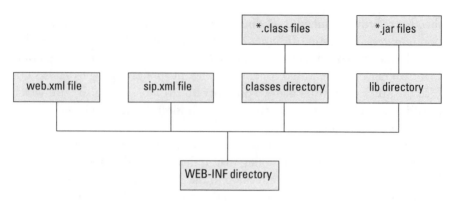

Figure 3.2   SIP and HTTP converged directory.

HTTP Servlet part by the inclusion of a "web.xml" file in the "WEB-INF" directory. On the flip side, an HTTP Servlet archive (".war" file) can contain a SIP Servlet part by the inclusion of a "sip.xml" file in the "WEB-INF" directory. The "WEB-INF" directory can then be included in a higher level Enterprise Archive (".ear" file) for inclusion in a larger JEE application.

3.1.1   Deployment Descriptor

The importance of the deployment descriptor file was certainly highlighted in the previous section. Its mere presence in the "WEB-INF" directory structure signifies SIP Servlet application components in an archive that might not actually be a SIP Servlet archive. The deployment descriptor is also used to specify a wide range of configuration values that are used in various forms by both the container and the application. This section will delve a little deeper to provide more information on the important elements that appear in a "sip.xml" file of the deployment descriptor. The top-level elements are named here, and appropriate child elements will be discussed in the relevant sections.

> *<app-name>*—Unique name of an application (".sar" file) within the context of a SIP Servlet container instance (including clustered container instances).
>
> *<distributable>*—Informs the container that the application being deployed is able to function correctly in a distributable environment.
>
> *<context-param>*—Allows an application to configure a set of initial application-level parameters that can be used during the life cycle.
>
> *<listener>*—The listener interfaces that are associated with the application.
>
> *<proxy-config>*—Configures proxy-related parameters.
>
> *<session-config>*—Session configuration as per JEE specifications.

<security-constraint>—Used to configure security constraints in association with SIP Servlets.

<login-config>—Enables an application to configure authentication mechanisms and appropriate parameters.

<security-role>—Security role configuration as per JEE specifications.

Figure 3.3 provides an example of a basic "sip.xml" deployment descriptor file containing some of the major elements discussed previously.

3.2 Application Roles

We have already discussed how, while the evolution of SIP Servlets has been very much influenced by the HTTP protocol and HTTP Servlet specification, a number of major differences exist. One of the main differences between HTTP and SIP is the nature of interactions. The HTTP protocol is very much a client–server protocol that results in simple request and response interactions. The SIP protocol, on the other hand, is based on peer-to-peer interactions that can result in either

```xml
<?xml version="1.0" encoding="UTF-8"?>
<sip-app xmlns:xsi="http://www.w3.org/2001/XMLSchema-instance"
xsi:schemaLocation=http://www.jcp.org/xml/ns/sipservlet
http://www.jcp.org/xml/ns/sipservlet/sip-app_1_1.xsd" ver-
sion="1.1">
<app-name>TestSample</app-name>
    <servlet>
        <servlet-name>TestSample</servlet-name>
        <servlet-class>net.sipservlet.sample.TestSample</servlet-
class>
    </servlet>
    <servlet-mapping>
        <servlet-name>TestSample</servlet-name>
        <pattern>
          <equal>
            <var>request.method</var>
            <value>INVITE</value>
          </equal>
        </pattern>
    </servlet-mapping>
</sip-app>
```

Figure 3.3 SIP Servlet deployment descriptor.

client issuing new request messages. The SIP protocol also employs the concept of multiple hops, which can result in a SIP message traversing multiple servers in its quest to find the destination. It is for this reason that the SIP Servlet architecture is based on a much varied set of functions that also heavily influence the SIP Servlet API and therefore the applications written. Taking the previous information into account, a number of roles can be assumed by applications to fully utilize the SIP protocol, including User Agent Client (UAC), User Agent Server (UAS), Proxy, and Back-to-Back User Agent (B2BUA). The roles are well-known constructs in the SIP protocol world and are mapped successfully to the SIP Servlet architecture. The following sections will provide an introduction to the roles. It should also be noted that the following roles within the general SIP architecture are normally independent entities. The SIP Servlet API allows an application to assume a specific role by key actions it takes on receiving a new SIP protocol request. In general, once an application has assumed a role for a request, changing it would violate core SIP principles. For this reason, a container will enforce specific rules on applications if they make decisions to assume a role but then later attempt an action specific to another role, which violates core SIP protocol and its associated state machines.

3.2.1 Proxy

A SIP Servlet application, using the SIP Servlet API (see Chapter 10 on the SIP Servlet API for more details related to using the API for the proxy role), is able to act as a SIP "Proxy." A SIP Proxy is defined in the core SIP Specification [1] as:

> An intermediary entity that acts as both a server and a client for the purpose of making requests on behalf of other clients. A Proxy server primarily plays the role of routing, which means its job is to ensure that a request is sent to another entity "closer" to the targeted user. Proxies are useful for enforcing policy (e.g., making sure a user is allowed to make a call). A proxy interprets, and, if necessary, rewrites specifics parts of a request message before forwarding it.

A SIP proxy server is an extremely powerful and widely used entity in SIP networks and provides the main routing engine for the protocol. It should be noted that a SIP Proxy has a strict set of rules that must not be violated in order to avoid undesirable consequences. As mentioned previously in this book, the very fact that the SIP protocol has the ability to locate and proxy a request onward to a more suitable location is one of the biggest differences between it and the HTTP protocol (and therefore the resulting API). HTTP Servlet containers are generally only required to generate responses to HTTP requests and have no concept of forwarding requests to a more appropriate location. Figure 3.4 illustrates a server that has taken an incoming SIP request and is now proxying it onward.

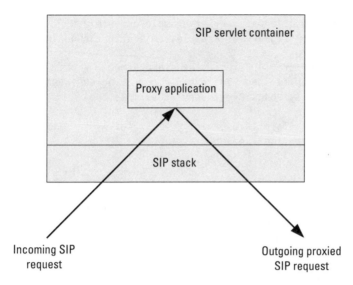

Figure 3.4   Simple proxy.

An incoming request that arrives at the application from the container is represented by a "SipServletRequest" interface. The application is able to initiate the intent to act as a proxy server by obtaining what is called the Proxy object from the "SipServletRequest" interface by calling as follows:

```
SipServletRequest.getProxy();
```

Now that the application has control of the Proxy object it has set its precedent to act in such a manner and can't act as one of the other roles discussed in this section. The application is able to use the Proxy interface to configure as appropriate for onward routing of the request. This might include, for example, adding multiple destinations to attempt when routing the request or searching through that list sequentially rather than in parallel. More detail on the configuration options for the Proxy interface can be found in the Appendix. Once the entire configuration for onward routing has been completed, the application calls the "proxyTo" method from the Proxy interface:

```
Proxy.proxyTo(sip:voicemail@sipservlet_example.com);
```

The SIP URI from the previous example would be used by the application to forward the request onward. For example, if the application received the following SIP INVITE request,

```
INVITE sip:kristoffer@sipservlet_example.com SIP/2.0
Via: SIP/2.0/UDP sipservlet_example.com;branch=z9hG483JKSJ8ew9
Max-Forwards: 70
To: Stoffe < kristoffer@sipservlet_example.com >
From: Chris <chris@sipservlet_example.com >;tag=8327489874
Call-ID: fj8493ijf984ulw94@sipservlet_example.com
CSeq: 1 INVITE
Contact: <sip:chris@pc.sipservlet_example.com >
Content-Type: application/sdp
Content-Length: 150
```

(Chris's SDP is not shown.)

The application would be presented with a "SipServletRequest" object by the container. It would use the ".getProxy" method on the object to obtain the Proxy object. By calling the previous ".proxy" example, it would forward the request to the SIP URI "sip:voicemail@sipservlet_example.com," and the outgoing SIP request would look like the following:

```
INVITE sip:voicemail@sipservlet_example.com SIP/2.0
Via: SIP/2.0/UDP proxy.sipservlet_example.com;branch=z9hG4a93uidal9
Via: SIP/2.0/UDP sipservlet_example.com;branch=z9hG483JKSJ8ew9
Max-Forwards: 70
To: Stoffe < kristoffer@sipservlet_example.com >
From: Chris <chris@sipservlet_example.com >;tag=8327489874
Call-ID: fj8493ijf984ulw94@sipservlet_example.com
CSeq: 1 INVITE
Contact: <sip:chris@pc.sipservlet_example.com >
Content-Type: application/sdp
Content-Length: 150
```

(Chris's SDP is not shown.)

Remember from the brief introduction to SIP at the beginning of this book that the top line of a SIP request, known as the Request URI (R-URI), indicates the next routing location. The R-URI in this example has been altered by the application from "sip:kristoffer@sipservlet_example.com" to "sip:voicemail@sipservlet_example.com." The simple code snippet for this application would look like the following:

```
public void doInvite(SipServletRequest local_Request)
{
        Proxy local_Proxy = local_Request.getProxy();
        local_Proxy.proxyTo(sip:voicemail@sipservlet_example.com);
}
```

An application can proxy to as many destinations as it wants, either by calling the "ProxyTo" method on the Proxy interface a number of times or by passing a preconfigured list as a parameter to the "proxyTo" method. The addition of more than one location to a proxy operation generates individual branches, as shown in Figure 3.5.

Using this mechanism for creating SIP proxy operations is certainly well understood and has been used extensively during the evolution of SIP Servlets. In the latest version of the specification, a more powerful variation for creating proxy operations has been introduced to complement the existing mechanism. As discussed earlier in this section, an application developer has the opportunity to configure various parts of a proxy operation before routing the SIP request onward. Unfortunately, such configuration values apply to all destinations used. For example, if we pass a list of five SIP URIs into a "proxyTo" method, then the configuration applies to all five SIP URIs. While this might be acceptable in most application scenarios, implementation experience has resulted in a more flexible approach that can be used, called Proxy Branches. Proxy Branches is similar to the "proxyTo" method, with the exception that it allows certain configuration values to change on specific branches. The mechanism for obtaining and configuring Proxy Branches is identical to that in the previous example, with the exception of its extra configuration flexibility. So, our previous code would look as follows:

```
public void doInvite(SipServletRequest local_request)
{
        Proxy local_proxy = local_request.getProxy();
        local_proxy.createProxyBranches(sip:voicemail@sipservlet_
example.com);
        local_proxy.startProxy();
}
```

This achieves exactly the same result as in the previous example of using the "proxyTo" method. It differs in that the application could call the "createProxy

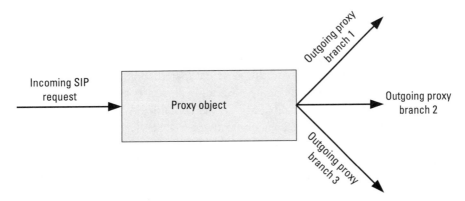

Figure 3.5 Proxy branches.

Branches" method again to add another location (note: you can call "proxyTo" method multiple times as well) until a final SIP response is passed upstream but with the important difference that the application developer can configure certain properties differently for each branch. The restricted properties that can be changed per branch are the following:

- Push alternate SIP route headers on each branch rather than the same one for all proxy locations.
- Issue a SIP CANCEL request on a specific branch rather than for all branches.
- Set different properties per branch on important SIP headers such as Record Route [1] and PATH headers [2].
- Set different recursive properties on each individual branch for SIP 300-class responses. In the SIP protocol, a 300-class response can either be acted upon by the Proxy or passed farther upstream.
- Specify a difference timeout value for each branch instead of for every branch.
- Generally manipulate SIP headers differently for each branch.

More detail on the Proxy Branches and Proxy interface API can be found in Chapter 10.

3.2.2   User Agent Client

A User Agent Client (UAC) is another role that a SIP Servlet application can assume and is defined by the core SIP specification [1] as follows:

> A user agent client is a logical entity that creates a new request, and then uses the client transaction state machinery to send it. The role of the UAC lasts for only for the duration of that transaction.

The UAC acts as a SIP client, and unlike a Proxy, it is an originator and sender of a SIP request from nothing, as compared to an entity that simply receives and forwards onward. Figure 3.6 simply illustrates how a UAC creates a request and sends externally from a SIP Servlet container. A Web-initiated call is a good example of UAC-type behavior in that a SIP Servlet container receives a request to start a call from a trigger that is not SIP (e.g., an EJB call). This would result in an application creating an appropriate SIP INVITE request and sending to the destination, therefore acting as a UAC. The application is, in effect, acting as a common SIP user agent, generating a call. Figure 3.6 provides a simple illustration of an application acting in the UAC role.

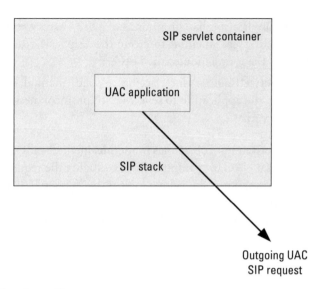

Figure 3.6 User Agent Client.

In the previous section, we discussed how a simple proxy server role can be carried out, and you might recall that in the simple code example the incoming SIP message was passed to the application in the form of a "SipServletRequest" object (from the "doInvite" interface method). An application acting as a UAC has a similar starting point. It is an originator of requests, does not have the incoming trigger of a proxy server, and is not passed a "SipServletRequest" object. It instead needs to be able to create a "SipServletRequest" object with no external SIP protocol trigger. It is clear that the container needs to provide a facility for applications to create a "SipServletRequest" object without an incoming SIP protocol trigger. Wait a second—that type of facility sounds familiar. Recall the example included in the section in Chapter 2 named "The SIP Servlet Container" that allowed the boss to be included in a conference call at a certain time using the previously discussed SIP Factory ("SipFactory" interface from SIP Servlet API) facility provided by the container? (If not, it would be worth going back to review that section.) So the SIP Factory interface provides the functionality we need to create the required trigger for a UAC-style application. In general, the following occurs when an application wants to act in the role of a UAC:

1. An instance of the SipFactory is obtained. This can be achieved either locally within the application, by obtaining a SIP Factory from the applications Servlet Context, or both locally and externally (from another JEE component like an EJB) by injecting an instance of the @SipFactory Java annotation (see examples of both techniques in the Chapter 2).

2. Once the application has an instance of a SIP Factory, it is able to use the "createRequest" method to create the "SipServletRequest" object that forms the foundation of the UAC.

3. The "SipServletRequest" interface has a "send" method, which is then invoked by the application to send the SIP protocol message, acting in the role of a UAC.

The "SipServletRequest" object has numerous method calls that enable an application acting as a UAC to manipulate and configure the request before it is sent. For further information, take a closer look at the "SipServletRequest" interface as detailed in the Appendix.

A simple code snippet for a UAC-based application generating a SIP INVITE request would look like the following:

```
@Resource SipFactory local_Factory_

SipApplicationSession appSession = local_Factory.createApplication
Session();
SipServletRequest local_Request = local_Factory.createRequest(app
Session, INVITE,
    sip:userA@sipservlet_example.com, sip:userB@sipservlet_example.
com);

[At this point the "SipServletRequest" request object would be con-
figured by the application using the appropriate methods and values:
String local_Sdp = . . . ]

local_Request.setContent(local_Sdp,"application/sdp");
local_Request.send();
```

First, an instance of the applications SIP Factory is injected into a local handle named "local_Factory." The factory is then used to create a "SipServletRequest" object called "local_Request." It should be noted that the SIP Factory "createRequest" method takes in a number of parameters that configure the local "SipServletRequest" object ("local_Request" from the example):

- The first parameter passed on specifies the SIP Application Session to which the request will be associated.
- The second specifies the type of SIP request we are generating: a SIP INVITE request in this case.
- The third parameter specifies who the request is from: the SIP URI "sip:userA@sipservlet_example.com" in our example. The value in this parameter would populate the SIP From header in the request.

- The fourth parameter specifies who the request is to—the SIP URI "sip:userB@sipservlet_example.com" in our example. The value of this parameter would populate the SIP "To" header and Request-URI in the request.

The new "SIPServletRequest" object created by such an operation also results in a new SIP Session (see later in this section for a detailed explanation of the relationship between SIP and a SIP Session). The SIP Session is also then associated with the Application Session specified in the "SipFactory" "createRequest" method call.

3.2.3 User Agent Server

A User Agent Server (UAS) is the third role in our series. After the discussion the role of the UAC in detail in the previous section the role of the UAS becomes obvious. The core SIP specification [1] defines the role of a UAS as follows:

> A user agent server is a logical entity that generates a response to a SIP request. The response accepts, rejects, or redirects the request. This role lasts only for the duration of that transaction.

So the role of UAS is, in fact, the exact opposite of a UAC. In the context of the SIP Servlet API, it is an application that is hosted on the container that makes a decision to not act as proxy (would not call the "getProxy" method on the SipServletRequest interface) when receiving a SIP request but generates a SIP response directly. Figure 3.7 provides an illustration of an application acting as a UAS.

In a similar fashion to the Proxy role, a UAS-based application is presented with a "SipServletRequest" object that is passed to it by the container. The "createResponse" method that appears on the "SipServletRequest" interface provides the ability for applications to act in the role of a UAS. On calling this method, the container will automatically generate a "SipServletResponse" object that is compliant to the SIP protocol based on the request object (in that all SIP headers are appropriately populated). The SIP response code used is specified as a parameter of the "createResponse" method. The application then has the opportunity to manipulate the request object as required using the other methods made available through the "SipServletResponse" interface. Once the application is happy that the "SipServletResponse" object is correct, it calls the "SipServletResponse" "send" method, which dispatches the response to the intended recipient.

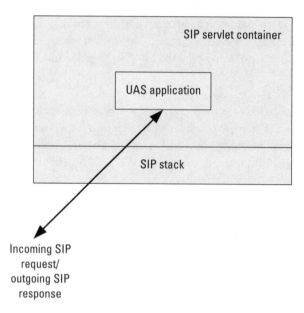

Figure 3.7 User Agent Server.

A simple example-code snippet from a UAS-based application receiving a SIP INVITE request would look like this:

```
public void doInvite(SipServletRequest local_Request)
{
        SipServletResponse local_Response = local_Request.create
Response(200);

        .... At this point the SipServletResponse object would be
configured by the
        application using the appropriate methods and values....

        local_Response.send()
}
```

The incoming SIP INVITE request is passed by the container to the application using the "doInvite" interface method. The application uses the "SipServlet Request" object (localRequest) to create a new SIP response object by calling the "createResponse" method on the "SipServletRequest" interface. Note that the application passes a "200" as a parameter to the method, which would result in a SIP 200 response being generated. The number would be changed appropriately for other responses. The application would now use other methods made available to manipulate the SIP response as required (additional information on functions available when manipulating the "SipServletResponse" object can be found in the section "SIP Servlet API"). Once the application is happy with the response that has been generated, it sends it to the intended recipient using the "SipServlet Response" "send" method.

3.2.4   Back-to-Back User Agent

The final role to be covered in this chapter is a Back-to-Back User Agent (B2BUA). A B2BUA is one of the most used constructs in SIP networks today and is defined as followed in the core SIP specification [1]:

> A back-to-back user agent (B2BUA) is a logical entity that receives a request and processes it in the user agent server. In order to determine how the request should be answered, it acts as a user agent client and generates requests. Unlike a proxy server, it maintains dialog state and must participate in all requests sent on the dialogs it has established.

In summary, a B2BUA in the context of the SIP Servlet API can be viewed as an application that acts as a concatenation of both the UAC and UAS roles that have been discussed in this section. In short, it is an application that is able to receive requests acting as a UAS, which then subsequently generates a request as a UAC for the same SIP protocol interaction. Figure 3.8 illustrates a simple B2BUA.

A B2BUA differs significantly from the Proxy role, which only acts on a SIP message to route it downstream. A B2BUA is responsible for generating a response (acting as a UAS) to the originator of the SIP request and also generating a SIP

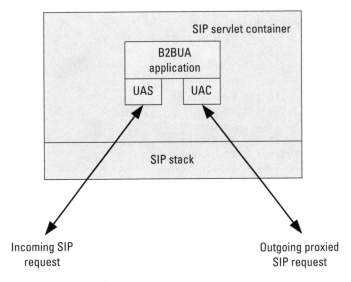

Figure 3.8   Back-to-Back User Agent.

request to route the call onward (acting as a UAC). The roles of UAC and UAS
are linked by the SIP Servlet API, which maps requests and responses between
both roles. The role of B2BUA is important for applications that need to maintain control over the SIP signaling, because SIP Proxy applications are not allowed
to generate SIP requests as per RFC 3261 [1]. A good example of a B2BUA would
be a prepay solution. It would receive calls and act as a UAS on one side while generating a new request to connect the call on the other side of the B2BUA. The
prepay application then starts a timer to check on the credit that is available to the
caller who initiated the call. When the credit timer fires, the application generates
the termination messages to hang up the call.

The B2BUA Helper only supports a one-to-one UA relationship. If dealing with a conference where the relation is one-to-many, it will not be of any help;
the various legs would have to be managed one by one. The following code provides an example of the B2BUA Helper class being used. The first Servlet is the
"B2BTerminatorServlet," which provides the main logic. If you work your way
through the code, you can see that, on first receiving a request (in the "doRequest"
method), an instance of the "B2BuaHelper" is obtained by calling the "getB2
buaHelper" method. The method is then eventually sent using the ".send" method
after copying the content of the message. In the "doResponse" method, the application retrieves the "B2buaHelper" instance and starts a timer on sending a SIP
"200 OK" response. After 60 seconds the call is terminated. Work your way through
the code to gain familiarity with how the SIP Servlet API is used to implement
this basic B2BUA. Functions are also included for dealing with ACK for messages
and the initial INVITE request. When the timer fires after 60 seconds, the "timeout" method is called, which results in the sending of the SIP BYE messages.

```
package net.sipservlet.sample;

import java.io.IOException;
import java.util.Iterator;
import java.util.List;
import java.util.logging.Level;
import java.util.logging.Logger;
import javax.annotation.Resource;
import javax.servlet.ServletException;
import javax.servlet.sip.B2buaHelper;
import javax.servlet.sip.ServletTimer;
import javax.servlet.sip.SipApplicationSession;
import javax.servlet.sip.SipServlet;
import javax.servlet.sip.SipServletMessage;
import javax.servlet.sip.SipServletRequest;
import javax.servlet.sip.SipServletResponse;
import javax.servlet.sip.SipSession;
import javax.servlet.sip.SipSessionEvent;
import javax.servlet.sip.SipSessionListener;
```

```java
import javax.servlet.sip.TimerListener;
import javax.servlet.sip.TimerService;
import javax.servlet.sip.UAMode;
import javax.servlet.sip.annotation.SipListener;

/**
 *
 * @author stoffe
 */
@SipListener
public class B2bTerminatorServlet extends SipServlet implements
TimerListener, SipSessionListener {

    @Resource
    TimerService ts;

    @Override
    protected void doRequest(SipServletRequest req) throws Servlet
Exception, IOException {
        if (req.isInitial() || req.getMethod().equals("ACK") ||
            req.getMethod().equals("CANCEL")) {
            super.doRequest(req); //Handled by the normal doXXX methods
        } else { //Subsequent signaling re-Invite, Bye
            B2buaHelper b2b = req.getB2buaHelper();
            SipSession linked = b2b.getLinkedSession(req.getSession());
            SipServletRequest other = b2b.createRequest(linked, req, null);
            copyContent(req, other);
            other.send();
            log("Subsequent request!" + req.getHeader("Cseq"));
        }
    }

    @Override
    protected void doResponse(SipServletResponse resp) throws Servlet
Exception, IOException {
        log("Got response : " + resp.getStatus());
        if (resp.getStatus() == SipServletResponse.SC_REQUEST_TERMINATED)
{
            return; //487 already sent on Cancel for initial leg UAS
        }

        B2buaHelper b2b = resp.getRequest().getB2buaHelper();
        SipSession linked = b2b.getLinkedSession(resp.getSession());
        SipServletResponse other = null;
        if (resp.getRequest().isInitial()) { // Handled separately due to
possibility of forking and multiple SIP 200 OK responses
            other = b2b.createResponseToOriginalRequest(linked, resp.get
Status(), resp.getReasonPhrase());
            //Start the timer on 200
```

```
      if (resp.getStatus() == SipServletResponse.SC_OK) {
        SipApplicationSession sas = resp.getApplicationSession();
        ServletTimer st = ts.createTimer(sas, 60000, false,
"enough"); //Cut the call after 60 sec
        sas.setAttribute("CallTimerId", st.getId());
        log("started the timer ID = " + st.getId());
      }
    } else { //Other responses then to initial request
      SipServletRequest otherReq = b2b.getLinkedSipServletRequest
(resp.getRequest());
      other = otherReq.createResponse(resp.getStatus(), resp.get
ReasonPhrase());
    }
    copyContent(resp, other);
    other.send();
    log("B2B Response " + resp.getHeader("Cseq"));
  }

  @Override //doInvite is only called for the first initial INVITE
  protected void doInvite(SipServletRequest req) throws Servlet
Exception, IOException {
    B2buaHelper b2b = req.getB2buaHelper();
    SipServletRequest other = b2b.createRequest(req, true, null);
    copyContent(req, other);
    other.send();
    log("Initial Invite! " + req.getHeader("Cseq"));
  }

  @Override //Only ACK for 200 error ACK created by container, find
it in pending
  protected

void doAck(SipServletRequest req) throws ServletException, IOException {
    log("Got ACK in.");
    B2buaHelper b2b = req.getB2buaHelper();
    SipSession ss = b2b.getLinkedSession(req.getSession());
    List<SipServletMessage> msgs = b2b.getPendingMessages(ss, UAMode.
UAC);
    for (SipServletMessage msg : msgs) {
      if (msg instanceof SipServletResponse) {
        SipServletResponse resp = (SipServletResponse) msg;
        if (resp.getStatus() == SipServletResponse.SC_OK) {
          SipServletRequest ack = resp.createAck();
          copyContent(req, ack);
          ack.send();
          log("Sent ACK out.");
        }
      }
    }
  }
```

```
@Override
protected void doCancel(SipServletRequest req) throws Servlet
Exception, IOException {
   log("Got CANCEL in.");
   B2buaHelper b2b = req.getB2buaHelper();
   SipSession ss = b2b.getLinkedSession(req.getSession());
   SipServletRequest cancel = b2b.createCancel(ss);
   cancel.send();
   log("Sent CANCEL out.");
}


private void copyContent(SipServletMessage source, SipServlet
Message dest) throws IOException {
   if (source.getContentLength() > 0) {
     dest.setContent(source.getContent(), source.getContentType());
     String enc = source.getCharacterEncoding();
     if (enc != null && enc.length() > 0) {
       dest.setCharacterEncoding(enc);
     }
   }
}


public void timeout(ServletTimer st) { //timer started on 200 for
Invite
   if ("enough".equals(st.getInfo())) { //terminate all B2B legs
     Iterator i = st.getApplicationSession().getSessions();
     while (i.hasNext()) {
       Object o = i.next();
       if (o instanceof SipSession) {
         SipSession ss = (SipSession) o;
         SipServletRequest bye = ss.createRequest("BYE");
         try {
           ss.setHandler("DummyResponseServlet"); //Not to send
response to the B2buaHelper
           bye.send();
         } catch (IOException ex) {
           log("", ex);
         } catch (ServletException ex) {
           log("", ex);
         }
         log("Sent bye to call-id : " + ss.getCallId());
       }
     }
   }
}


public void sessionCreated(SipSessionEvent event) {
}
```

```
public void sessionDestroyed(SipSessionEvent event) {
}


public void sessionReadyToInvalidate(SipSessionEvent event) { //BYE
state, cancel the timer
    SipApplicationSession sas = event.getSession().getApplication
Session();
    String timerId = sas.getAttribute("CallTimerId").toString();
    ServletTimer st = sas.getTimer(timerId);
    if (st != null) { //Already canceled, we forced the BYE
      st.cancel();
      log("canceled the timer ID = " + timerId);
    }
  }
}
```

The second Servlet, "DummyResponseServlet," is a simple function included to log the receipt of a response to the SIP BYE requests.

```
package net.sipservlet.sample;

/**
*
* @author stoffe
*/
public class DummyResponseServlet extends javax.servlet.sip.
SipServlet {

  @Override
  protected void doResponse(javax.servlet.sip.SipServletResponse
resp) throws javax.servlet.ServletException, java.io.IOException {
    log("Got response for BYE : "+resp.getStatus());
  }
}
```

3.3 Application Constructs

In the scope of an application, there are two well-known primary constructs that are used and that should be familiar to SIP Servlet application developers. They are the SIP Application Session and SIP Session. Both of these concepts have been touched upon during other chapters of this book, and this section aims to provide a detailed reference.

The reader is already well aware of the tight evolutionary relationship that the SIP Servlet specification has with the HTTP Servlet specification. This theme continues when introducing both SIP Session and SIP Application Session concepts. Protocols such as HTTP and SIP can involve a series of messages that are related and need to be processed as such. In HTTP this is solved by using the concepts of a "HttpSession," which represents the association of multiple numbers of HTTP requests. The SIP Servlet architecture has a similar concept in the form of SIP Sessions, which are used to correlate a series of messages from and to a user. As SIP Servlets are stateless when deployed, the SIP Session is also used to store specific information relating to the SIP signaling interaction. The differing types of interactions that can occur in the SIP protocol can result in the need for many protocol sessions to be associated. The SIP Application Session provides an umbrella construct that allows an application to achieve such correlation. The SIP Application Session represents a specific instance of an application, which in turn can contain zero or more associated SIP Sessions representing SIP protocol signaling interactions, as Figure 3.9 illustrates.

In the same way that, due to the statelessness of SIP Servlets, an application developer can store related data in the SIP Session, there is also the ability to store application data in the SIP Application Session. More detail is provided on these two extremely important concepts in the next two sections.

3.3.1 SIP Application Session

A SIP Application Session is a fairly simple construct within the SIP Servlet API (compared to SIP Session) and primarily has two functions. It provides the developer the ability to store application-related data persistently across both protocol and nonprotocol invocations. It also acts as a correlating structure for multiple associated SIP signaling interactions. A conference is a good example of a representation of a SIP Application Session correlating a number of SIP protocol sessions (SIP Sessions): An employee decides to schedule a conference call and include ten

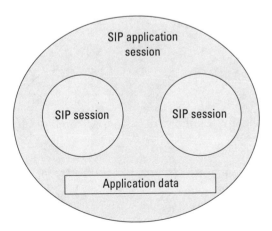

Figure 3.9 SIP Application and SIP Session.

employees from the company, including himself. The conference call is due to start at 4 p.m. that day. When 4 p.m. arrives, the SIP Servlet-based application creates a SIP Application Session to represent the employee's conference call. As time goes on, a number of invited attendees successfully dial into the conference call as instructed by the host. As each of the attendees dials in, it creates a one-to-one SIP signaling interaction with the Application Session representing the conference. By the time everyone has dialed in, we are left with a single SIP Application Session that contains 10 associated SIP Sessions, as shown in Figure 3.10.

3.3.2   SIP Session

A SIP Session represents SIP protocol sessions that are being managed by the SIP Servlet container. RFC 3261 [1] defines a SIP dialog as being "…a peer-to-peer SIP relationship between two UAs that persist over some time."

It is, therefore, obvious that the state machine for a SIP Session is almost identical to that of a SIP dialog. When the SIP Servlet container receives a SIP request and passes it to the appropriate application, a SIP Session is created to represent that protocol interaction between two entities.

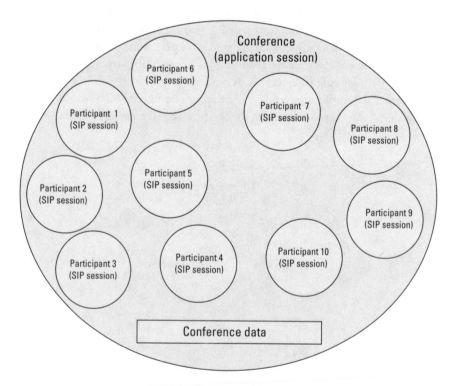

Figure 3.10   SIP Application Session conference example.

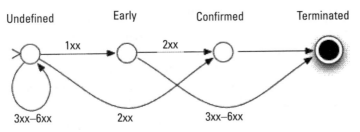

Figure 3.11 SIP dialog state machine.

Figure 3.11 is taken from the SIP Servlet specification [3] and illustrates the various states of a SIP dialog. Figure 3.12 is also taken from the SIP Servlet specification [3] and illustrates the state machine of a SIP Session.

The main difference between the two state machines can be seen as an additional state, called "INITIAL," that was introduced in the SIP Session state machine (as described later in this section).

The state machine represented by a SIP Session is extremely important due to the close relationship with the SIP protocol. Application developers are able to carry out a wide, powerful range of protocol operations that could violate the SIP Session state machine and cause undesired protocol signaling side effects. It is for this reason that an application developer should act responsibly when carrying out various protocol-level operations. A container must also enforce the state models discussed in this section by throwing an "IllegalStateException" when a violation occurs.

As noted from Figure 3.12, a SIP dialog has four possible states that it can be in, depending on the state of the associated SIP signaling protocol session. The states are as follows:

- *Initial*—Represents the additional state introduced in the SIP Servlet API when no SIP dialog is currently in progress (e.g., before you receive a SIP protocol response and after you receive a negative response). The initial state allows for an application to carry out further processing on

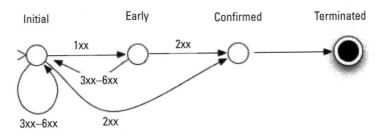

Figure 3.12 SIP Session state machine.

receipt of an unsuccessful SIP protocol response. In SIP, the receipt of such negative responses results in the termination of a SIP dialog. In SIP Servlet programming, it is often useful to be able to carry out subsequent operations that might result in the resending of the SIP request, which might then be successful. The addition of the INITIAL state allows for such flexibility in SIP Servlet-based applications.

- *Early*—A SIP Session is in the early state when it receives a SIP provisional response in the range of 101 to 199.

- *Confirmed*—A SIP Session is in the confirmed state when it receives a 200 class response.

- *Terminated*—In some scenarios, it is not possible for the state to transition back to the INITIAL state, because it would break the SIP state machine.

The SIP Servlet specification has a strict set of rules that govern how a container should transition to and from the SIP Session states depending on the type of role it occupies (as in UAC, UAS, or Proxy application). In general, non-dialog-creating requests do not result in a state change from INITIAL. Non-dialog-creating requests are considered single interactions that do not expect to have subsequent messages sent or received. The SIP MESSAGE [4] method can be seen as an example of a nondialog-creating request. It represents a one-shot, SMS-style message between users. An INVITE request, however, does represent a SIP dialog-creating request that maps to the full state machine for SIP Session.

An application acting as a UAC and sending a dialog-creating request such as INVITE would use the full state machine discussed in this section to map to the SIP dialog interactions, with the exception on receiving a non-200-class error response. While in the INITAL or EARLY state, the SIP Session state would transition back to INITIAL rather than TERMINATED. If the SIP Session state had transitioned to TERMINATED, then the SIP Session would have been over. Transitioning back to the INITIAL state gives the application developer the opportunity to attempt the request again, maybe with a change to the signaling as a result of the response or to route to an alternative location that might be successful.

An application assuming the role of a UAS that is handling a dialog-creating request also tracks the SIP Session but does not have the exception of transitioning back to the INITIAL state when issuing a non-200 class response. It in fact moves to the TERMINATED state, because the interaction is complete, and no further processing can take place when acting in this type of role.

Applications acting as a Proxy have a role similar to the role of a UAC in that they map the SIP Session state as per Figure 3.12, but they transition back to the INITIAL state on receiving a non-200 class SIP response. As with a UAC, this allows an application developer to attempt alternative SIP signaling to create an established dialog. This is most common in popular applications, where the result of a failure leads to an alternative location being attempted, such as directing a call to voicemail when the user does not answer. Such flexibility is essential to make SIP Servlet programming useful when acting as a UAC or Proxy. If the proxy server does not include such logic to catch a non-200 class response, then it is considered the best response and is passed upstream to the originating client. The selection of the best non-200 class response by SIP proxy servers to be passed upstream is a well-known SIP operation defined in RFC 3261 [1]. On sending the non-200 class response upstream, the SIP Session state would transition to the TERMINATED state. If the best response being passed upstream was a 200 class response, then the SIP Session state would transition to CONFIRMED.

3.3.3 Application Data Storage

SIP Application Sessions and SIP Sessions, which have been previously introduced in this section, provide containers for application data storage. SIP Servlets are considered stateless in that they do not directly monitor or hold any specific transactional data. When dealing with a number of correlated SIP messages in association with an application, it is the responsibility of the application developer to decide what application data needs to be stored across multiple SIP transactions.

The SIP Servlet API provides convenience mechanisms on both SIP and SIP Application Sessions. The following description applies to both. An application has the ability to bind object attributes to SIP and SIP Application Sessions. An object that an application binds to either the SIP or SIP Application Session is then accessible to any other Servlet that makes up the application archive (".sar" file). Just to reiterate, this does not mean that other SIP Servlet applications can gain access to the attributes stored—only SIP Servlets that fall under a specific Servlet Context (boundary within a ".sar" file). Figure 3.13 illustrates application data being stored.

SIP Session and SIP Application Session interfaces provide the following methods to enable applications to bind, manage, and remove attributes:

* *"SipSession.setAttribute"* and *"SipApplicationSession.setAttribute"*—An application is able to bind a Java object representation of the application data to either the SIP or SIP Application Session using this method call. It simply passes in the object as a parameter along with a unique name of type 'string' for identification purposes. It is important for an application to devise an appropriate naming convention when adding new attributes to a SIP or SIP Application Session. Adding an attribute with the same name as one that already exists will result in the previous value being overwritten.

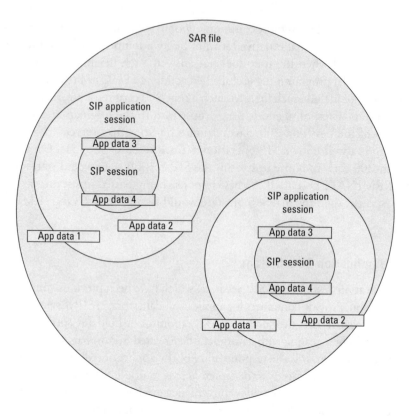

Figure 3.13   Application data storage.

- *"SipSession.getAttribute"* and *"SipApplicationSession.getAttribute"*—As the name suggests, this method call allows an application to retrieve a previously stored Java object that had been bound using the "setAttribute" method on the SIP and SIP Application Session objects. The method call simply takes the unique string name of the attribute as a parameter and returns the previously stored Java object.

- *"SipSession.getAttributeNames"* and *"SipApplicationSession.getAttribute Names"*—Calling this method, which takes no parameters, on either a SIP or SIP Application Session results in a list of unique string names being returned for valid attributes that have been bound previously.

- *"SipSession.removeAttribute"* and *"SipApplicationSession.removeAttribute"*— An application is able to remove or unbind data by calling the "remove" attribute method on either SIP or SIP Application Session interfaces and including the unique attribute string value as a parameter in the call to identify the appropriate object.

The SIP Servlet API provides all the tools to easily manage SIP and SIP Application Session data across an application. It is important to note that any number of components in a SIP Servlet application could rely on and also have the ability to manage the bound application data (e.g., add, remove, and change). For this reason, the SIP Servlet API has listener interfaces for both SIP Session and SIP Application Session. The "SipApplicationSessionBindingListener" and "SipSessionBindingListener" both can be implemented by applications that want to receive notifications on attribute data that is bound. An application implementing either of these interfaces will receive a notification when an object is bound or unbound to the SIP or SIP Application Session. It receives a "SipSession Binding" event, which can then be used to identify the unique name associated with the application data attribute and also to obtain the associated SIP or SIP Application Session, if required.

3.3.4   Session Lifetime and Invalidation

SIP and SIP Application Sessions have emerged as the major constructs that make up the SIP Servlet architecture. It is therefore important to ensure they are managed appropriately by both SIP Servlet containers and applications. A division of responsibility exists with the life-cycle of such session objects with a strong emphasis that SIP and SIP Application Session objects are efficiently managed. The following sections will cover the life-cycle and invalidation of SIP and SIP Application Sessions.

3.3.4.1   SIP Application Session

A SIP Application Session can be created from a number of mechanisms that have been discussed in this book. For example, a SIP Application Session could be created as a result of an incoming SIP protocol message or from using the container's SIP Factory utility. The amount of time that a SIP Application Session is considered active is specified in two ways:

1. It can be specified using the "session-timeout" parameter that exists in the application deployment descriptor ("sip.xml").
2. It can also be specified using the "@SipApplication" (sessionTimeout) Java annotation for configuring "SipServlet" applications.

The value contained in both the deployment descriptor and the Java annotation indicates a value in seconds that the SIP Application should be considered "alive" for application processing. If neither of these two mechanisms is specified, the timeout value for a SIP Application Session defaults to three minutes. If either the deployment descriptor or @SipApplication did specify a value that was either negative or zero, the timer should not be started, and the container should consider the SIP Application Session to be long lived and thus necessarily managed by the application. Managing the SIP Application Session wholly within the Application logic does work but must be heavily guarded so as not to leave around redundant SIP Application Session objects that will never be cleaned up and will therefore cause memory leaks.

The SIP Servlet API defines a "SipApplicationListener" interface that supplies appropriate information to applications relating to the SIP Application Session. One of the methods called on the interface is "sessionExpired," which provides application logic with the SIP Application Session object that has just expired. The application logic then has the ability to request an extension to the lifetime of the SIP Application Session object by calling the "setExpires" method on the SIP Application Session interface. The container will return a value based on the request from the "setExpires" method call and will either accept an extension by returning a positive number of seconds (note that the returned time may not be the one requested by the application and could be smaller, for example, if local policy dictates) or return the value of zero to indicate that the application to extend the SIP Application Session lifetime was rejected. If the application does not call the "setExpires" method within the implementation of the "Sip ApplicationListener" interface (or doesn't even implement the "SipApplication-Listener" interface) then the SIP Application Session is considered expired and ready to be appropriately invalidated. The invalidation of an expired SIP Application Session results in the destruction of the object so that it can no longer be used, including any bound application data. The container should also call the "sessionDestroyed" method that appears on the "SipApplicationSessionListener" interface.

Historically, SIP Application Sessions were simply explicitly invalidated by the application when it was considered appropriate. This led to a number of problems and race conditions in which developers were invalidating sessions incorrectly, which often led to complicated code structures to determine when invalidation should take place. The latest version of the SIP Servlet architecture provides a much cleaner, complementary mechanism that uses container knowledge to cleanup SIP Application Sessions only when appropriate. The two mechanisms are known as "Explicit Invalidation" and "Invalidate When Ready."

Explicit Invalidation—This basically involves the application logic calling the "invalidate" method that appears on the "SipApplicationSession" interface. This results in the purging of the SIP Application Session and all its related data with immediate effect and does not take into consideration the state of the object or its related protocol sessions. This is the original SIP Servlet mechanism for invalidation and should be used with caution.

Invalidate When Ready—An elegant approach to invalidation that takes into account the associated SIP protocol sessions and timer objects that might be active. A SIP Application Session transitions from an active to "invalidate-when-ready" state when the final associated protocol session has been invalidated and the last timer object has expired. In order to complement this mechanism, the following new methods were added in the latest version of the SIP Servlet architecture to the "SipApplication Session" interface:

isReadyToInvalidate—Returns a Boolean value that is set to "true" if the SIP Application Session is in the "ready-to-invalidate" state and "false" otherwise.

setInvalidateWhenReady—A container will monitor a SIP Application Session and notify the container using the "sessionReadyToInvalidate" method on the "SipApplicationListener" callback interface. An application can use this "setInvalidateWhenReady" Boolean value to indicate whether the container should notify using this method. A value of "true" instructs the container to call the "sessionReadyToInvalidate" method on the "SipApplicationListener" interface, while "false" instructs the container not to call this method. For applications written to the latest SIP Servlet specification (based on version 1.1), the default value is "true," and for older applications (based on version 1.0) the default value is "false."

getInvalidateWhenReady—This method returns a Boolean value indicating the status of the container in relation to its monitoring state. So a value of "true" is returned if it is currently monitoring for the "ready-to-invalidate" state, and "false" is returned if it is not monitoring.

As mentioned in the previous list, if the container is instructed to notify the application when a SIP Application Session moves into the "ready-to-invalidate" state, it must call the "isReadyToInvalidate" method on the "SipApplication Listener" interface. On receiving such a notification, the application could choose to invalidate the session using the previously described explicit invalidation mechanism by calling the "invalidate" method on the "SipApplicationSession" interface. The difference this time is that the application is confident that the SIP Application Session object is in a state that constitutes a safe invalidation. The application can also carry out any other tasks that might be required at this appropriate juncture. It might be that the application does not explicitly invalidate the Sip Application Session in the code or in fact it has not even implemented the "Sip ApplicationSessionListener" interface. In these cases, the container will automatically invalidate the session at the appropriate time. In the case that an application has implemented the "SipApplicationSessionListener" but has not explicitly invalidated, this occurs when the "isReadyToInvalidate" method returns from the application.

3.3.4.2 SIP Session

A SIP protocol session or SIP Session mirrors its parent SIP Application Session lifetime and invalidation mechanisms. A SIP Session does not have its own specified timeout values and naturally inherits the value for the SIP Application Session. If a SIP Application Session transitions to the expired state, it results in all associated SIP Sessions also transitioning to the inherited expired state.

The SIP Session has the same invalidation mechanisms as a SIP Application Session, with the additional inheritance invalidation as a result of the parent SIP Application Session's being invalidated.

> *Explicit Invalidation*—As with the SIP Application Session, this involves the original invalidation mechanism specified by the SIP Servlet architecture, where application logic can call the "invalidate" method on the "SipSession" interface. This results in the SIP Session object and all its related application data's being destroyed immediately. The explicit invalidation of SIP Sessions has the same problems as its SIP Application Session parent in that, if the underlying protocol session is not in the correct state to be invalidated, it can cause spurious and negative message exchanges. The application code would need complex code structures to ensure that inappropriate invalidations and race conditions don't occur.

> *Invalidate When Ready*—This also follows a pattern similar to its SIP Application Session equivalent but with some subtle differences to map to the underlying protocol session semantics. A SIP Session should be considered "ready to invalidate" only when the underlying SIP protocol interactions are in a completed state. This SIP Session state is monitored by the container and only considered transitioned to the "ready-to-invalidate" state, which is defined in the SIP Servlet specification [3] thus:

> - A dialog corresponding to a "SipSession" terminates when the "SipSession" transitions to the TERMINATED state.

> - A "SipSession" transitions to the CONFIRMED state when it is acting as a nonrecord-routing proxy.

> - A "SipSession" acting as a UAC transitions from the EARLY state back to the INITIAL state on account of receiving a non-2xx final response and has not initiated any new requests (does not have any pending transactions).

The "SipSession" interface defines the same three method calls as the "SipApplicationSession" interface but with slightly different semantics:

isReadyToInvalidate—Provides an application with a Boolean return value to indicate whether the container is monitoring the SIP Session and will notify an interested application when it has transitioned to the "ready-to-invalidate" state. A value of "true" indicating that the SIP Session is ready to invalidate and a value of "false" indicating that it is not ready to be invalidated.

setInvalidateWhenReady—This method allows an application to programmatically instruct the container to monitor a SIP Session and inform an interested application when it has transitioned to the "ready-to-invalidate" state. A value of "true" tells the container to report "invalidate-when-ready" state, and a value of "false" tells the container not to report "invalidate-when-ready" state. If the value is set to true, the container will invoke the "sessionReadyToInvalidate" method on the "SipSession Listener" callback interface when the SIP Session transitions to the "invalidate-when-ready" state. The application would need to implement the "SipSessionListener" to receive such notifications.

getInvalidateWhenReady—Provides a Boolean return value that indicates to an application whether the container is monitoring "invalidate-when-ready" status on a given SIP Session. A value of "true" informs the application that "invalidate-when-ready" state for the SIP Session is being monitored by the container, while a value of "false" informs the application that "invalidate-when-ready" state for the SIP Session is not being monitored by the container. The default value indicating the "invalidate-when-ready" status of SIP Sessions is set to "true" for applications using the latest version of the SIP Servlet specification (Version 1.1) and "false" for those applications using an older version (Version 1.0).

Parent Invalidation—As previously mentioned, SIP Sessions are totally dependent on their parent SIP Application Session. If the parent SIP Application Session is invalidated, for example, using the explicit invalidation mechanism, the underlying SIP Sessions will be destroyed automatically, along with any associated application data.

An application wishing to safely invalidate a SIP Session should use the "invalidate-when-ready" mechanism discussed in this section. This is achieved by implementing the "SipSessionListener" interface and making use of the "session ReadyToInvalidate" method that the container calls when the SIP Session transitions into the "ready-to-invalidate" state. On receiving a call from the container to this method, the application can then safely invoke the explicit invalidation mechanism using the "invalidate" method on the "SipSession" interface. It can also then clean up any other application state. If the application does not use explicit invalidation within the implementation of the "SipSessionListener" (or the application does not even implement the interface), the container will automatically invalidate the SIP Session and clean up any related application data.

At the SIP protocol level, it is possible that a container might receive SIP protocol signal after a SIP or SIP Application Session has been invalidated. Depending on local policy, the container should choose to either attempt "best effort" routing to try to complete the transaction—but without application intervention—or, if policy does not allow, an appropriate SIP error response such as "481" should be returned.

3.3.5 Annotations

We have already discussed how Java annotations play a major role in the evolution of SIP Servlet architecture as it moves toward JEE 5 integration. This section will take a closer look at some of the Java annotations that should be supported by a container and used in various ways by applications. Some annotations that provide container-injected utilities such as SIP Factory, Session Timer, and Session Utilities have already been discussed, while others are included to reduce the need for static deployment descriptor files. The following is a selected list of important Java annotations that are used within applications:

@SipServlet—This top-level annotation actually reduces the need for a deployment descriptor to exist at all. Its presence alone in a class defines a SIP Servlet application. The "@SIPServlet" annotation then has a number of elements that are used to represent certain parameters that appear in the deployment descriptor being used to specify a SIP Servlet application. These elements include the following:

- *name*—Represents the "servlet-name" name element from the deployment descriptor file.
- *applicationName*—Specifies the associated application name.
- *description*—Provides information relating to the application.
- *loadOnStartup*—Maps to the "loadOnStartup" element that can appear in the deployment descriptor.

@SipApplication—This annotation is used represent an application that can consist of multiple SIP Servlets, as defined by the @SipServlet annotation. This annotation also has a number of elements that are used to provide the appropriate configuration that could have appeared in a deployment descriptor. These elements include the following:

- *name*—Represents the "application-name" element from the deployment descriptor and is mandatory for the latest versions of the specification (Version 1.1).
- displayName—Represents the "display-name" element of the deployment descriptor, providing an appropriate application display name.

- *largeIcon*—Represents optional "large-icon" element from the deployment descriptor.

- *smallIcon*—Represents optional "small-icon" element from the deployment descriptor.

- *description*—Represents optional "description" element from the deployment descriptor that is intended to provide contextual information about the function of the application.

- *distributable*—Represents the "distributable" attribute from the deployment descriptor that signifies if the application is distributable across platforms.

- *proxyTimeout*—Represents the "proxy-timeout" element from the deployment descriptor that specifies a general timeout figure in seconds for SIP protocol proxy transactions.

- *sessionTimeout*—Represents the "session-timeout" element from the deployment descriptor that specifies in minutes the timeout period for SIP Application Sessions.

- *mainServlet*—Represents "main-servlet" element from the deployment descriptor that indicates which Servlet is considered the default for initial SIP incoming requests.

@SipListener—This annotation provides an alternative to the "listener" element from the deployment descriptor and annotates a class to be a listener. The listener classes being used as Java indicated by the implementation of the class.

@SipApplicationKey—Used to help associate new incoming requests with existing SIP Application Sessions. Use of this annotation for session targeting is discussed Chapter 10.

SipFactory Injection—Uses the Java @Resource annotation to inject an instance of the SIP Factory into an application. This is discussed in detail in Chapter 2.

SipSessionUtil Injection—Uses the Java @Resource annotation to inject an instance of the SIP Session Utilities into an application. This is discussed in detail in Chapter 2.

TimerService Injection—Uses the Java @Resource annotation to inject an instance of the Timer Service into an application. This is discussed in detail in Chapter 2.

For those familiar with earlier versions of SIP Servlet programming (Version 1.0), the latest version of the specification provides a table that indicates how deployment descriptor properties can now be mapped to the to various Java @ annotations. The table, taken from JSR 289 [3], looks as in Figure 3.14:

Display Name	`display-name`	`@SipApplication displayName`
Description	`description`	`@SipApplication description`
Distributable	`distributable`	`@SipApplication distributable`
Context parameters	`context-parm ,` `param-name ,` `param-value`	Not applicable
Listener	`listener-class`	`@SipListener`
Servlet name	`servlet-name`	`@SipServlet name`
Application name	`app-name`	`@SipApplication name`
Servlet class	`servlet-class`	`@SipServlet`
Initialization parameters	`init-parm ,` `param-name ,` `param-value`	Not applicable, suggested to use constants.
Startup order	`load-on-startup`	`@SipServlet loadOnStartup`
Proxy timeouts	`proxy-timeout`	`@SipApplication proxyTimeout`
Session timeouts	`session-timeout`	`@SipApplication sessionTimeout`
resources	`resource-*`	`@Resource , @Resources`
Security Roles	`security-role *`	`@DeclaresRole`
EJBs	`ejb-ref *`	`@EJB`
Run as	`run-as *`	`@RunAs`
Web Services	Not applicable	`@WebServiceRef`

Figure 3.14 Annotations mapping.

The core Servlet specification, from which the SIP Servlet specification derives, requires that a number of annotations be supported. The reader should refer to Servlet 2.4 (see [5] to view the complete list and usages of supported annotations). As well as supporting the annotations included in Servlet 2.5, a SIP Servlet container has the following list of additional classes and interfaces that must be supported for compliance in conjunction with Java annotation support:

- Javax.servlet.sip.SipServlet Servlet class interface;
- Javax.servlet.sip.SipApplicationSessionListener listener interface;
- Javax.servlet.sip.SipApplicationSessionActivationListener listener interface;
- Javax.servlet.sip.SipSessionAttributeListener listener interface;
- Javax.servlet.sip.SipSessionListener listener interface;

- Javax.servlet.sip.SipSessionActivationListener listener interface;
- Javax.servlet.sip.SipErrorListener listener interface;
- Javax.servlet.sip.TimerListener listener interface.

The version of SIP Servlet specification (Version 1.1—JSR 289) that introduces Java annotation support into the architecture takes a subset from Servlet 2.5, the common Java annotations specification [6], and adds newly created SIP Servlet-specific annotations. The complete list of Java annotations that are supported by a JEE-compliant container, as defined by the SIP Servlet architecture, can be listed as follows:

@RunAs
@DeclaresRole
@Resource
@Resources
@EJB
@WebServiceRef
@PostConstruct
@PreDestroy
@SipServlet
@SipApplication
@SipListener
@SipApplicationKey

The topic of various deployment architectures was covered earlier in the book. Those deployment architectures that are not part of a JEE system are required to support only Java annotations that are related to SIP.

References

[1]   Rosenberg, J., et al., "SIP: Session Initiation Protocol," RFC 3261, Internet Engineering Task Force, June 2002.

[2]   Willis, D., and B. Hoeneisen, "Session Initiation Protocol (SIP) Extension Header Field for Registering Non-Adjacent Contacts," RFC 3327, Internet Engineering Task Force, December 2002.

[3]   SIP Servlet Specification, Version 1.1, JSR 289, Java Community Process, August 2008.

[4]   Campbell, B., et al., "Session Initiation Protocol (SIP) Extension for Instant Messaging," RFC 3428, Internet Engineering Task Force, December 2002.

[5]   Java Servlet Specification, Version 2.5, JSR 154, Java Community Process, September 2007.

[6]   Common Annotations for the Java Platform, JSR 250, Java Community Process, May 2006.

4

Application Router

SIP Servlet technology had evolved dramatically since its conception and has taken advantage of a large amount of industry interest and support. Such adoption has resulted in an evolving technology that has kept pace with increasing requirements and adapted to fulfill key roles in various next-generation architectures. One of the primary advancements between SIP Servlet 1.0 (JSR 116) and SIP Servlet 1.1 (JSR 289) was the introduction of a logical role called the Application Router. The Application Router aids in providing a new, sophisticated application composition model that is highly configurable and adaptable. This section will provide a detailed overview of the new application composition model and the Application Router.

4.1   SIP Servlet 1.1 Composition Model

SIP Servlet 1.1 introduced a powerful capability for allowing multiple applications to be invoked by an initial SIP signaling request on a single visit to a SIP Servlet container. As described in Chapter 3, an application archive (".sar" file) contained a deployment descriptor file (sip.xml) in the same way HTTP Servlets have a configuration file (web.xml). A SIP Servlet container, on receiving a new SIP request, inspects the <servlet-mapping/> element of the deployment descriptor XML file, which provides the rule set for requests to be passed into a SIP Servlet application. An example of a <servlet-mapping/> element is as follows:

```
<servlet-mapping>
    <servlet-name>ExampleServlet</servlet-name>
    <pattern>
```

```
        <equal>
         <var>request.method</var>
         <value>INVITE</value>
        </equal>
     </pattern>
   </servlet-mapping>
```

This provides a simplistic view of a <servlet-mapping/> element that informs the SIP Servlet container to pass all initial SIP INVITE requests to the application associated with the "sip.xml" deployment descriptor file. The container will move on to the next application and its associated "sip.xml" deployment descriptor file in a preconfigured order that is implementation specific. The <servlet-mapping/> can also makes use of "and" and "or" structures that allow multiple combinations of entry configuration to be set. The following provides a simple "or" operation when checking a new request:

```
<servlet-mapping>
   <servlet-name>ExampleServlet</servlet-name>
     <pattern>
        <or>
           <equal>
                 <var>request.method</var>
                 <value>INVITE</value>
           </equal>
           <equal>
             <var>request.method</var>
             <value>SUBSCRIBE</value>
           </equal>
        </or>
     </pattern>
   </servlet-mapping>
```

In this example, a container will pass a new SIP request to the SIP Servlet named "ExampleServlet" only if the message is either an INVITE or a SUBSCRIBE. In a similar manner, the "and" operator can be used as follows:

```
<servlet-mapping>
   <servlet-name>ExampleServlet</servlet-name>
     <pattern>
        <and>
           <equal>
                 <var>request.method</var>
                 <value>INVITE</value>
           </equal>
           <equal>
             <var>request.to.uri.user</var>
```

```
            <value>chris</value>
        </equal>
     </and>
  </pattern>
</servlet-mapping>
```

In this example, a container will pass a new SIP request to the SIP Servlet named "ExampleServlet" only if the message is of type INVITE and the user part (before "@") of the SIP URI in the "To" header is equal to the string "chris." The "and" and "or" operators can also be combined (as well as other constructs) to start building complex filtering rules, as shown in the next example:

```
<servlet-name>ExampleServlet</servlet-name>
<pattern>
    <and>
    <or>
       <equal>
          <var>request.method</var>
          <value>MESSAGE</value>
       </equal>
       <equal>
          <var>request.method</var>
          <value>REFER</value>
       </equal>
    </or>
    <or>
       <equal>
          <var>request.from.uri.user</var>
          <value>Simon</value>
       </equal>
       <equal>
          <var>request.from.uri.user</var>
          <value>Chris</value>
       </equal>
          <not>
             <equal>
                <var>request.from.uri.user</var>
                <value>Kristoffer</value>
             </equal>
          </not>
       </or>
    </and>
</pattern>
</servlet-mapping>
```

Although it is a contrived example, it illustrates an outer "and" block with multiple embedded "or" blocks. In short (in pseudo code),

```
IF the SIP request method equals "MESSAGE" OR "REFER"
AND
The user in the SIP "From" header equals "Simon," "Chris," or
"Kristoffer," then pass to the application;
```

It should be noted that more than one SIP Servlet class can exist within a "sip.xml" and can be represented by a <servlet-mapping/> element. The <servlet-name/> element specifies a pointer to a SIP Servlet class and can be included multiple times when creating more complex application admission rules. The following example conveys multiple options for a container when attempting to dispatch a SIP request to a hosted application:

```
<servlet-mapping>
    <servlet-name>INVITEServlet</servlet-name>
    <pattern>
      <equal>
       <var>request.method</var>
       <value>INVITE</value>
      </equal>
    </pattern>
  </servlet-mapping>
<servlet-mapping>
    <servlet-name>SUBSCRIBEServlet</servlet-name>
    <pattern>
      <equal>
       <var>request.method</var>
       <value>SUBSCRIBE</value>
      </equal>
    </pattern>
  </servlet-mapping>
```

In this example, two separate Servlets exist within a single SIP Servlet archive (".sar" file). SIP INVITE requests will be passed to the "INVITEServlet," while SIP SUBSCRIBE requests will be passed to the "SUBSCRIBEServlet." A "sip.xml" can have any number of Servlets. If more than one Servlet matches, generally it's the first (and only the first) one that receives the request.

For more information on the options available when using the <servlet-mapping/> element of a SIP Servlet deployment descriptor, please see the appropriate XML schema provided as part of the SIP Servlet specification.

Figure 4.1 provides an illustrative view of a SIP request traversing a container using SIP Servlet 1.0 application composition.

On the left-hand side of Figure 4.1, there is an arrow that signifies an incoming SIP request into the SIP Servlet container. Once the request has been parsed by the SIP stack, it is passed to the container as a new SIP request. This is denoted by the "Decision Point" on Figure 4.1. The container inspects the

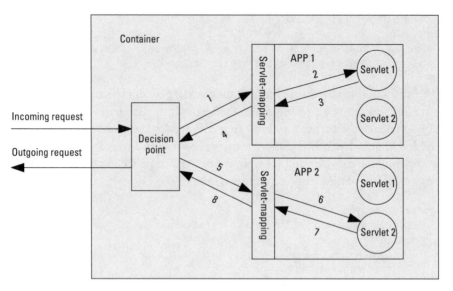

Figure 4.1 SIP Servlet 1.0 application routing.

elements of the SIP Servlet applications ("<.sar" file), which are deployed in a predefined configuration order. On finding a match (as discussed earlier in the introduction to the XML format), the container dispatches the request to "App 1," as illustrated by (1). The Servlet mappings dictate which of the SIP Servlet classes will receive the incoming request. At (2), the element routes the request to "Servlet 1." On finishing with the request, "Servlet 1" returns it back the container to make the next decision, as indicated by (3) and (4). The container then dispatches the request to "App2," (5). The elements are again inspected, which results in the request being dispatched to "Servlet 2," (6). Once processing of this request has completed, the request is returned back to the container, (7) and (8). The container then identifies that no more SIP Servlet applications should service the SIP request. The SIP request is then routed onward on its journey, according to core SIP protocol rules as defined in RFC 3261 [1].

This example illustrates a straightforward approach to application composition and routing, which changes dramatically when you consider that an application can assume different roles. For example, an application acting as a User Agent Server (UAS) would not dispatch the request back to the container once it had completed processing but would generate a SIP response message that acts as the final destination for the SIP request. Similarly, an application acting as a User Agent Client (UAC) would generate its own, unsolicited SIP requests based on a third-party trigger. The general rules for assessing and dispatching requests remain the same, just with different behavior (mostly focused on entry and exit of the SIP request) depending on role.

This brief introduction to SIP Servlet 1.0 application composition and routing touches on the power that can be achieved within a SIP Servlet container. For several years, this technique provided adequate application composition and routing functionality. It is fair to say that in the SIP Servlet 1.0 specification the use of this mechanism from a container perspective was not that clearly defined. While all containers could easily comply with the appropriate deployment descriptor XML files, it was found that container behavior in this area varied between implementations. It was also noted that, while the static XML style filtering that was used in SIP Servlet 1.1 allowed for a consistent selection process, it was not ideal as more complex application composition scenarios were emerging.

As a result of such experiences, a new application composition model is introduced in SIP Servlet 1.1 that builds on the foundations of previous experience. Whereas in SIP Servlet 1.0 there was reliance on the container inspecting the deployment descriptor XML files for the purpose of dispatching a SIP request to an application, SIP Servlet 1.1 includes an explicit application selection mechanism. A logical entity called an "Application Router" is introduced to act as the decision point for application selection in a composition chain. Figure 4.2 illustrates the new application selection process, in which a container asks the "Application Router" to inform it of the next application selection step.

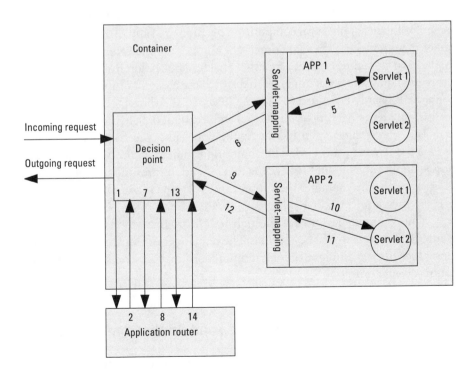

Figure 4.2 SIP Servlet 1.1 application routing.

The initial step is identical to that described in the previous SIP Servlet 1.0 example. As illustrated on the left-hand side of Figure 4.2, a new SIP request arrives at the SIP Servlet container. Once it has been passed to the container, it reaches a point where the container must decide the next step. This is where the process differs. The container now uses the new logical entity called the Application Router. In simplistic terms, the container is asking the Application Router which application it should dispatch to next. This takes the onus away from the container and puts the responsibility on the Application Router.

In the example in Figure 4.2, the container has asked the Application Router where it should send the new SIP request (1). The Application Router inspects the request and determines that the request should be routed to "App 1." The Application Router responds to the container, informing it to dispatch the request to "App1," at (2) in the figure. The container then dispatches the new SIP request to "App 1," at (3). It should be noted that the use of the <servlet-mapping/> element is now no longer required in this composition model (which will be discussed later). It is a well-known construct that is used widely by SIP Servlet application developers and is now an optional part of an application that can be used as a final filtering rule to allow SIP requests entry. The favored method is to use the "main Servlet" approach in conjunction with the Request Dispatcher and Handler as discussed in Chapter 2. The SIP request reaches "App 1," indicated by (4) in the figure, and application logic is executed until the request is returned back to the container for further application composition decisions, (5) and (6). At (7), the container then carries out the same process of asking the Application Router which application should receive the request next. The Application Router responds to the request, informing the container to dispatch the request to "App 2," at (8). The container forwards the request to "App 2," (9) and (10), which processes the request (carries out appropriate application logic) and returns it back to the container, (11) and (12). The container again asks the Application Router which application should receive the request next, (13). The application router determines that no more hosted applications should receive the request and informs the container appropriately, (14). The container then routes the SIP request as defined by the core SIP specification RFC 3261 [1].

This example has provided a high-level introduction of the Application Router and its role in SIP Servlet 1.1. The remainder of this chapter will take a closer look at the main interactions and functions that are involved in the application routing process.

4.2 Application Router, Container, and Application Interaction

The Application Router provides a powerful decision-making entity that removes ambiguity and provides clarity in the application selection and composition process of SIP Servlet 1.1. It has always been the intention that an Application Router not be tied in any way to the container with which it operates and the applications that it is selecting. In reality, the majority of Application Routers will be produced by the associated container vendors, but allowing independence provides a lot more flexibility and power. The majority of concepts discussed in this chapter provide such independence.

In order to fulfill such requirements, it is important that Application Router implementations are transferable across varying container implementations for differing vendors. It is the container's responsibility to identify and instantiate the Application Router when it starts up. To achieve this, the SIP Servlet 1.1 specification mandates that container implementations be packaged to comply with the JAVA SE Service Provider Framework. As part of this mandate, an Application Router ".jar" file has to contain a file named "SipApplicationRouterProvider." The full path for the file is "META-INF/services/javax.servlet.sip.ar.spi.SipApplication RouterProvider." The content of this file allows the container to identify the public Java subclass of the "javax.servlet.sip.ar.spi.SipApplicationRouterProvider." SIP Servlet 1.1 gives more detail on how a container is able to identify the Application Router instance and instantiate the correct version.

Once an Application Router has been successfully deployed by a container, it is available for service on receiving new SIP requests. For Application Routers to be truly portable requires a simple, common interface that containers can call to learn of the next application to route to (as illustrated in the early SIP Servlet 1.1 composition example). The "SipApplicationRouter" interface, which is part of the SIP Servlet API, provides the required common interaction point and must be implemented by all Application Routers. On commencement of service, it is important that the container has a mechanism for informing the Application Router which applications are available for service. This allows the Application Router to make informed decisions based on the applications available rather than blindly route SIP requests. The "SipApplicationRouter" interface has two methods that provide such a service. The container calls the "SipApplicationRouter.application Deployed" method, which provides the Application Router with a list (in the form "java.util.List<java.lang.String>") of new applications that have been deployed. The container continues to call this method during the life-cycle of the container to inform of newly deployed applications. On the flip side, an Application Router also needs to be made aware when applications are undeployed and taken out of service. The "SipApplicationRouter.ApplicationUndeployed" method is called by the container during its life cycle, supplying a list of (type "java.util.List <java.lang.String>") applications that have been removed from service. This allows the Application Router to ensure it is making decisions on applications that are definitely available for service.

The other important method in the "SipApplicationRouter" interface provides the key interaction for a container when it wants to dispatch a SIP request to an application instance that is currently in service. The "SipApplicationRouter. getNextApplicationRouter" method is repeatedly called by container for a request until no further applications are intended to service the SIP request. The method provides the appropriate context to an Application Router to allow for an appropriate application selection decision to be made. The return value from the method call not only informs the container of the appropriate application to dispatch to but also supplies appropriate context for future invocations relating to the same SIP request. By supplying the container with such contextual information allows an Application Router to remain stateless, which improves efficiency and performance.

Referring back to the example provided in Figure 4.2, on receiving an initial request the container executes the "SipApplicationRouter.getNextApplication" method to inquire about dispatching of the SIP request. It is the container's role to populate this method call with enough information so that the Application Router can make an educated decision. The method has the following parameters:

- The first parameter is an instance of the "SipServletRequest" interface, which represents the incoming SIP request. The Application Router is not allowed to modify in any way the interface instance representing the SIP request, because its intention is to provide information and context to aid the application selection process.

- The second parameter is of the type "SipApplicationRoutingRegion" interface, which basically provides the Application Router with a string of "NEUTRAL_REGION," "ORIGINATING_REGION," or "TERMINATING_REGION." The concept of the routing region is quite straightforward. If a request is being serviced with applications on behalf of a user that generated a SIP request, the container and Application Router are acting in the originating region. The user that is "originating" can be determined from various mechanisms, such as Digest Authentication, as defined in RFC 3261 [1], and SIP Identity [2]. If no authentication mechanism is available, then the SIP "From" header could simply be used. Once a container has finished servicing an originator of a SIP request, it carries out the same procedures but this time sets the parameter as "TERMINATING_REGION." The terminating region signifies the intended recipient of the SIP request as specified by the SIP Request URI. This continues until the Application Router returns a value of "null" in response to a request for the next application, which then results in the SIP request being routed as per core SIP protocol. There also exists a third routing region that can optionally be executed in the middle of the originating and terminating regions. The "Neutral" region allows the container to execute applications on a SIP request that are not really associated with the originator or terminator of a SIP request. A simple example might be a logging application that is required to log all events regardless of the routing region. To recap, on receiving an initial SIP request, the

container will first invoke the "SipApplicationRouter.getNexApplication
Router" method with the routing region set to "ORIGINATING_
REGION" until the Application Router has executed all appropriate
applications. It will then optionally repeatedly call the "SipApplication
Router.getNextApplication" method with the routing region set to
"NEUTRAL_REGION" until the Application Router has executed all
appropriate applications. It will then repeatedly call the "SipApplication
Rotuer.getNextApplication" method with the routing region set to
"TERMINATING_REGION" until the Application Router returns
"null" as the name of the next application. Figure 4.3 provides a graphical representation.

It should noted that the previous description makes an assumption
that the SIP Servlet container is the primary application service function
in the deployment (services all users). This most definitely will not always
be the case, and it is more than likely that users will be spread across
multiple instances of SIP Servlet containers that will also span geographical boundaries. It is not always the case, therefore, that a SIP
Servlet container will carry out originating processing and terminating
processing on a new SIP request. Often, a "home" instance of a SIP
Servlet container will carry out the originating application processing
for a user before using SIP to route the request to the "home" instance
of the SIP Servlet container providing terminating application services
for the intended recipient of the SIP request. Figure 4.4 provides a graphical illustration of such an arrangement.

- The third parameter in the method call provides an Enum contact signifying the routing directive of the request. The routing directive is a concept introduced in SIP Servlet 1.1. The directive can have one of three
  values that relate to how the Application Router should treat the request
  in the context of a composition chain. The following lists the context that
  each routing directive provides:
  - *NEW*—The container informs the Application Router that the
    request is brand new and is not part of an existing application

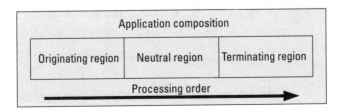

Figure 4.3 Region processing.

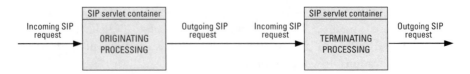

Figure 4.4 Split application composition.

composition chain. The Application Router should treat it as such and select the first application to service the appropriate user.

○ *CONTINUE*—The container informs the Application Router that this request is part of an existing application composition chain that has been seen previously. The request has already been dispatched to at least one application, so the appropriate next application should be selected in the composition chain.

○ *REVERSE*—This directive is not as common as the previous two. The container informs the Application Router to reverse the direction of the call. The SIP Servlet 1.1 specification provides cases in which this feature can be useful.

The routing directive of a request obtains its values in one of two ways. By default it is derived by the actions taken by applications and then supplied to the Application Router. The implicit deductions for the routing directive are dependent on the role an application is assuming (User Agent Client, User Agent Server, or Proxy), based on the following SIP Servlet API method calls:

○ If the SIP factory is used (as introduced in Chapter 2) to create a request based on an instance of the "SipApplicationSession" interface when acting as a User Agent Client, the container will assume a "NEW" directive. For example:

```
Sip_Request = Sip_Factory.createRequest(SipApp
Session, "INVITE," sip:from@example.com,
sip:to@example.com)
```

○ If an application uses the "Proxy" interface, it assumes the role of a SIP proxy server. The container will automatically set the routing directive to "CONTINUE" in this case, as in the following:

```
Request.getProxy().proxyTo()
```

○ If the SIP Factory is used (as introduced in Chapter 2) to create a request based on an instance of the "SipServletRequest" interface when acting as a User Agent Client, the container will assume a

"CONTINUE" directive. This differs from the previous SIP Factory example in that the new request is based on an incoming request, which implies a Back-to-Back User Agent (B2BUA). The previous Factory example was not based on an incoming request and so is assumed to be unrelated to any other requests, as in the following:

```
Sip_Request = B2buaHelper.createRequest(SipServlet
Request)
```

○ While the default behavior has been discussed, there are cases in which an application intentionally does not want a directive to be assumed. The "SipServletRequest.setRoutingDirective" method of the SIP Servlet API allows an application to override the default container behavior and set the routing directive to one of the valid values, as follows:

```
Sip_Request.setRoutingDirective(CONTINUE,
orig_SipServletRequest)
```

Full details of the "SipServletRequest.setRoutingDirective" method can be found in the Appendix.

- The fourth parameter is used when an incoming request is determined to be a targeted request. A targeted request is defined as one that contains either a SIP "Join" header or a SIP "Replaces" header, or one whose Request URI contains an encoded URI (as defined in the SIP Servlet 1.1 specification). The parameter is an instance of the "SipTargetedRequestInfo" class, which informs the Application Router of the type of targeted request it is dealing with (a SIP "Join" header, SIP "Replaces," or an Encode URI). Before calling the Application Router, the container inspects the list of active "SipSession" interface instances to see whether a match can be found for the targeted operation in the new SIP request. If an instance is found, the corresponding "SipApplicationSession" instance and application name is determined. If the container deduces that the identified "SipApplicationSession" instance is acting in the role of a User Agent (either User Agent Client or User Agent Server) and not a Proxy, then the application name is included in the instance of the "SipTargetedRequestInfo" class along with the target type. If the request is not a targeted request or the "SipApplicationSession" can't be located or the located instance was acting as a Proxy, the "SipTargeted RequestInfo" is set to "null."
- The fifth parameter is an object of type "java.io.Serializable," which allows an Application Router to store appropriate state information related to

the composition chain. On initial requests, the state object is set to "null," but on future calls to the Application Router the state object that is returned as part of this method call must be included. This enables the Application Router to store information that, along with the other parameters in this method call, allow for an accurate application selection in a composition chain. It also allows the Application Router to remain stateless.

On receiving the "SipApplicationRouter.getNextApplication" method call, the Application Router can extract all the supplied information as well as the state information if it is a subsequent call for a SIP request. At this point, the mechanisms and logic used by an Application Router to decide on the next application for the SIP request (if one at all) is totally implementation specific. The interface only provides a simple front end to interact with the container. The outcome of the application selection process could be as simple as looking at a static file and as complex as required, maybe even including interactions with third-party servers [such as a database or Home Subscriber Server (HSS) query]. Once the Application Router has made its decision, it composes its response to the "SipApplication Router.getNextApplication" method call. This is accomplished by populating an instance of the "SipApplicationRouterInfo" class, which contains the following methods for the container to use:

1. The first method (SipApplicationRouterInfo.getNextApplicationName) provides a string representation of the application name being hosted by the container that the request should be dispatched to.

2. The second method (SipApplicationRouterInfo.getRouteModifier) returns contextual information related to the third method call (SipApplicationRouterInfo.getRoutes), which will be discussed next. A value of "NO_ROUTE" indicates to the container that the "SipApplication RouterInfo.getRoutes" method does not contain any valid routing information. A value of "ROUTE" indicates to the container that the "Sip ApplicationRouterInfo.getRoutes" method will return valid SIP Route headers that should be populated in the SIP request (as the top-most SIP "Route" headers). A value of "ROUTE_BACK" instructs the container to insert a SIP "Route" header pointing to itself before including the SIP "Route" headers that are returned as a result of "SipApplication RouterInfo.getRoutes." This has powerful connotations when in certain distributed environments where applications can be hosted on a variety of disparate server instances. As an example, the IP Multimedia Subsystem (IMS) architecture has a centralized application composition entity, called the Serving Call Session Control Function (S-CSCF). This entity uses the pushing of SIP "Route" headers (pushing a SIP "Route"

header that points back to itself), which allows a SIP request to return once it has been serviced by an application. Further processing and application selection can then take place. The ability for a SIP Servlet container to select applications that are hosted either locally or on an external server is an extremely powerful tool that can be used when designing complex application composition chains.

3. The third method (SipApplicationRouterInfo.getRoutes) returns an array of type string that the container must add to the front of the SIP route set. These SIP "Route" headers define the next routing steps for the request. For more information relating to the use of pushing preloaded SIP "Route" headers see the core SIP specification [1]. It should be noted that a SIP "Route" header can be both internal and external to the SIP container. It is not mandatory for an Application Router to push a SIP "Route" header when routing to an internal application, because the application name returned as part of this method call provides enough information to request SIP routing. It is an optional approach that can be viewed as a more complete, elegant solution that truly treats hosted applications as independent SIP entities. Effectively, an internal SIP "Route" header that appears in this method call would be pushed and popped (added to and removed to the list of valid destinations) immediately. This is certainly a purist approach that provides true application independence, because in theory, it could be hosted anywhere, and not specifically on the container instance in question.

4. The fourth method call (SipApplicationRouterInfo.getRoutingRegion) provides the routing region that this SIP request is currently being executed in. The previously discussed values of the routing region can be "ORIGINATING_REGION," "TERMINATING_REGION," and "NEUTRAL_REGION." It is then the container's responsibility to store the appropriate region associated with a SIP request so it can be supplied back the Application Router on further queries (invocations of the Sip ApplicationRouterInfo.getNextApplication) for the same SIP request. This helps in keeping the Application Router as lightweight and stateless as possible.

5. The fifth method (SipApplicationRouterInfo.getSubscriberURI) informs the container of the subscriber that is being serviced. This differs for a SIP request depending on the routing region (either originating or terminating) being executed.

6. The sixth and final method (SipApplicationRouterInfo.getStateInfo) enables the container to obtain the associated state with the application composition chain. The container will store this instance of "java.io. Serializable" and provide it back to the Application Router on further

invocations of the "ApplicationRouter.getNextApplication" method for the same SIP request. This allows the Application Router to store any proprietary, relevant information that will enable it to accurately resume the composition chain. This supports the premise that an Application Router can be totally stateless.

It should be noted that full, detailed descriptions of the SIP Servlet API interfaces, methods, and class related to this area are provided in Chapter 10.

It would be useful to take the previously discussed information and relate it back to the example SIP request flow that was introduced in Figure 4.2. Please refer back to that figure in relation to the following description: A new request arrives at the container, at which point a decision needs to be made on exactly which applications should receive the request. The container would call the previously described "SipApplicationRouter.getNextApplication" method (1), populating the appropriate parameters. This includes the following:

- A read-only copy of the SIP request;
- The routing region (probably set to "ORIGINATING_REGION" in the example);
- The routing directive (would be set to "NEW" on an initial request);
- Set to "null" because it is not a targeted request;
- State information also set to "null" because it is a new request.

On receiving the request from the container, the Application Router would populate the appropriate parameters in the "SipApplicationRouterInfo" object to return back to the container, (2). For the purpose of the example, this would include the following:

- The name of the application that should receive the request;
- The routing region (still set to "ORIGINATING_REGION" in the example);
- The Subscriber URI (set to the originating user for the request, as determined by authentication, for example);
- An empty list of SIP "Route" headers;
- A route modifier with a value of "NO_ROUTE";
- An object containing proprietary state information.

The container inspects the instance of the "SipApplicationRouterInfo" and carries out a number of tasks. It first stores appropriate information returned by the Application Router that will enable it to continue the application composition chain in the future. This includes storing the routing region and the state information object. The container then looks at the string returned by the "Sip ApplicationRouterInfo.getApplicationName" method, which would have returned "App 1." The container then dispatches the SIP request to "App 1," (3). The Application completes its processing and, acting as a SIP proxy entity, forwards the message downstream, (5) and (6). On receiving the request back, the container identifies that it has seen this request before and retrieves the previously stored information. The container then populates the next interaction with Application Router, using the "SipApplicationRouter.getNextApplication," which would look as follows:

- A read-only copy of the SIP request;
- The routing region (probably set to "TERMINATING_REGION" in the example);
- The routing directive (would be set to "CONTINUE" for this subsequent request as application acted as a Proxy);
- Set to "null" because this is not a targeted request;
- State information, returned previously by the Application Router, included because this is a continuation of a composition chain.

On receiving the request, (7) in the figure, the Application Router inspects the supplied information and notices the "CONTINUE" routing directive, which identifies that this is a continuation of a sequencing chain. By using the other appropriate information, including the proprietary state information object, the Application Router is able select and dispatch to the appropriate application ("App 2") and return that back to the container, at (8), for further processing, (9) and (10). The rest of the example follows the same pattern. This is a rather simplistic view and provides good context of the interactions that take place among a SIP Servlet container, an Application Router, and an application.

4.2.1 Subsequent Requests and Responses

The focus of this chapter so far has been on the composition of initial SIP requests that arrive at a SIP Servlet container. Once a SIP request has been routed successfully through a container, it has usually visited a number of applications. For example, in Figure 4.5 the SIP request traverses three applications.

The traversal of applications creates what is known as an application path, so in Figure 4.5, the application path equals App1 $\rightarrow$ App2 $\rightarrow$ App3. A consequence of this path's being created internally to the container is that SIP responses associated with the original request should traverse the same applications but in the opposite direction. In our previous example, a SIP response would follow the application path App3 $\rightarrow$ App2 $\rightarrow$ App1. This is illustrated in Figure 4.6.

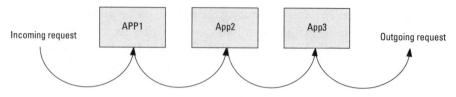

Figure 4.5   SIP request application path.

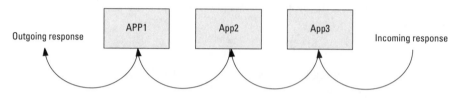

Figure 4.6   SIP response application path.

Dialog-creating requests in SIP, such as INVITE/REFER/SUBSCRIBE, create a long-lived association between two endpoints (for more information on SIP dialog-creating requests, see the core SIP specification [1] and Chapter 1). As a result, subsequent, associated SIP messages can traverse in either direction, and they should follow the same application path as the original request when a container acts as either a User Agent or a record-routing proxy server. The order in which the application path for subsequent requests is traversed is based entirely on the directionality of the subsequent request. If the endpoint generating the subsequent request destined for the container is the same as the originator of the initial request, then the application path would be the same as the original request: App1 → App2 → App3 for our earlier example. The application path is illustrated in Figure 4.7.

If the originator of a subsequent request is on the receiving end of the original request (the endpoint that previously generated the SIP response to the initial request), the application path is reversed in the same way as for the generated SIP response, for example, App3 → App2 → App1. The application path is illustrated in Figure 4.8.

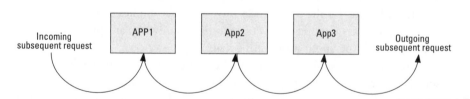

Figure 4.7   SIP subsequent request application path from originator.

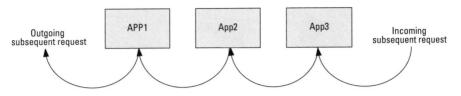

Figure 4.8 SIP subsequent request application path from receiver.

This chapter has provided a brief introduction to the logical role of an Application Router and the glue that binds it with a SIP Servlet container and SIP Servlet applications. The business logic behind an application is implementation specific and can be as simple or complex as required. The only requirement is that the basic interactions that have just been covered are adhered to by the applications, SIP Servlet containers, and Application Routers.

References

[1]   Rosenberg, J., et al., "SIP: Session Initiation Protocol," RFC 3261, Internet Engineering Task Force, June 2002.

[2]   Peterson, J., and C. Jennings, "Enhancements for Authenticated Identity Management in the Session Initiation Protocol (SIP)," RFC 4474, Internet Engineering Task Force, August 2006.

5

Moving Forward

Voice-over IP technology has come a long way since its conception in the late 1990s. New frontiers and boundaries are being pushed on a regular basis as industries converge on new multimedia communications that take advantage of ever-improving network infrastructure. To date, SIP Servlet technology has evolved from its conception, SIP Servlet 1.0, to the latest release, SIP Servlet 1.1. This demonstrates a commitment from the industry to ensure that the requirements that are being spawned from ever-evolving technology are met. Failure to synchronize the technology with its main customers will result in a technology that depreciates in value as time progresses. On building the requirements phase for SIP Servlet 1.1, it was obvious that a number of topics were far too vast and complicated to be included. For this reason, an extensive exercise was undertaken to determine, first, what could adequately be covered in the short time scales available and, second, what was appropriate to include in a minor release of the technology. This process was then tempered by the fact that SIP Servlet technology needs to remain aligned with the ever-evolving landscape of the communications industry. The success of SIP Servlet technology is evident in the number of commercial products available in the market place as well as high profile, publicly reported live deployments. This is especially impressive when you consider the usual trepidation that such new technologies are greeted with in the industry. This widespread adoption and implementation experience forms the basis for current and any future advancements in SIP Servlet technology.

5.1 SIP Servlet Threading Model

The evolution of SIP Servlet 1.1 very much leads the technology toward integration with Java Enterprise Edition (JEE) and inclusion as part of larger, potentially distributed applications. In such environments, multiple applications will require access to SIP Servlet session objects at the same time (e.g., access to the "SipApplicationSession" interface).

SIP Servlet 1.1 is silent on specifying a threading mechanism that should be used by containers when dealing with multiple threads' access to objects such as instances of the "SipApplicationSession" and "SipSession" interfaces. It recommends that the responsibility lie with the application developer to ensure synchronization with these interfaces and their associated application data. It also mentions that containers could provide a proprietary thread-safe mechanism that ensures such session objects are accessed appropriately.

The SIP Servlet 1.1 expert group considered including an optional thread-safe mechanism for containers to implement. The mechanism would have included an API and associated listener to enable applications to schedule works to be carried out by the container. It was the intention that the container would then schedule and execute specified tasks without danger of competing for access to session objects. The application would then use the associated listener interface to obtain notifications on completion of the scheduled work.

It was decided by the SIP Servlet 1.1 expert group that introducing a thread-safe mechanism would be a step too far for a relatively minor release of the specification. The next release of the technology, which more than likely will be a major 2.0 release, will include an appropriate access mechanism "SipApplicationSession" and "SipSession" objects.

5.2 Outstanding Issues

It is guaranteed that the implementation and deployment experience gained with SIP Servlet 1.1 will lead to a new range of issues and changes to be addressed by the next release of the technology. It should be mentioned that the list of new requirements logged for the SIP Servlet 1.1 version was extensive, and due to time constraints not all could be addressed. The list of remaining requirements will form the basis for any future versions of SIP Servlet technology.

5.3 SIP Protocol Support

SIP Servlet technology obviously has a tight association with the core SIP protocol and its main extensions that are defined by the Internet Engineering Task Force (IETF). Some of the primary SIP extensions have been incorporated into the technology to provide increased coverage of the protocol. Examples include SIP Reliable Responses [1], SIP "Join" header [2], and SIP "Replaces" header [3]. It is important that future releases also track important evolutions in the core SIP protocol and provide support to aid application developers. Major topics such as Network Address Translation (NAT) traversal (and its associated solutions), support for core protocol changes, and other important protocol evolutions should be considered for inclusion.

5.4   JSR 309

A new and evolving Java Specification Request (JSR) that will provide a media services API for controlling media servers is currently under development. As illustrated in Figure 5.1 (taken from the JSR 309 documentation), the complementary relationship of the two technologies provides an extremely powerful solution.

JSR 309 implementations are intended to take full advantage of SIP Servlet technology depending on which of the underlying media server protocols they are using. The options range from MSML [4] to MSCML [5] to MediaCtrl [6], and so on. The exact integration and deployment models surrounding JSR 309 and SIP Servlets are yet to be established and will become apparent as JSR 309 matures and as more implementation experience is gained. SIP Servlet application developers should feed such experiences into the SIP Servlet process to ensure tight technology associations for an optimal media services solution.

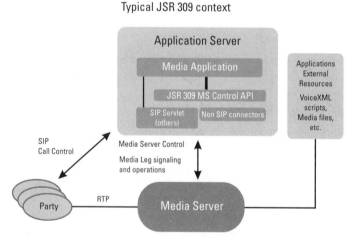

Figure 5.1   JSR 309 context.

References

[1]   Rosenberg, J., and H. Schulzrinne, "Reliability of Provisional Responses," RFC 3262, Internet Engineering Task Force, June 2002.

[2]   Mahy, R., and D. Petrie, "The Session Initiation Protocol (SIP) 'Join' Header," RFC 3911, Internet Engineering Task Force, October 2004.

[3]   Mahy, R., B. Biggs, and R. Dean, "The Session Initiation Protocol (SIP) 'Replaces' Header," RFC 3891, Internet Engineering Task Force, September 2004.

[4]   Saleem, A., Y. Yin, and G. Sharret, "Media Server Markup Language (MSML)," draft-saleem-msml, Internet Engineering Task Force.

[5]   Van Dyke, J., E. Burger, and A. Spitzer, "Media Server Control Markup Language (MSCML)," RFC 4722, Internet Engineering Task Force, November 2006.

[6]   Media Server Control (MediaCtrl), www.ietf.org/html.charters/mediactrl-charter.html. Internet Engineering Task Force.

Part II
Developer and Deployment Environments

6

Relationship and Role Within IMS

The IP Multimedia Subsystem (IMS) is an architecture that provides a framework for next-generation IP networks. It is defined by the Third Generation Partnership Project (3GPP) in Europe and 3GPP2 in North America (and Japan and South Korea). More recently, the Telecoms and Internet Converged Services and Protocols for Advances Networks (TISPAN) group was spawned to ensure representation for fixed-line networks (i.e., signaling and media carried across a physical communication technology) as well in mobile IMS (i.e., signaling and media carried across an appropriate air interface). The design aims is for an access agnostic architecture that allows for consistent service delivery for both wireless and wireline technologies, potentially running in parallel. It promotes strong principles like security and quality of service while aiding convergence of fixed and wireless access mechanisms. The IMS architecture is based primarily on IETF protocols (mainly SIP for signaling), which enables interoperation of a wide variety of third-party entities. 3GPP has defined some additional extensions through the IETF for IMS, and although it does slightly tweak certain aspects of the protocol, the protocol is mainly untouched.

A major focus of the IMS architecture is the ability to rapidly introduce new services to a network based on a wide variety of Application Server types. Figure 6.1 provides a simple view representing a generic application server's role in the IMS architecture.

Without delving too much into the detail surrounding the IMS architecture (there are dedicated resources describing this large technological area), one of the primary entities involved is called the Serving Call Session Control Function (S-CSCF), which manages users and their applications. This is achieved by handling SIP REGISTER transactions, which map the availability of an IMS user as

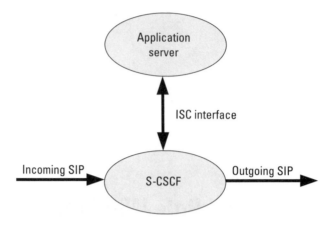

Figure 6.1   ISC interface.

well as look up the appropriate applications (using appropriate interfaces to IMS data storage) that should be invoked for a certain SIP request type. The requests are then dispatched on the IP Multimedia Subsystem Service Control (ISC), which is a SIP-based interface that must be implemented by Application Servers wishing to integrate into the IMS architecture. Once an Application Server has completed its service-related processing, the SIP request is returned back to the S-CSCF for further downstream processing. This could either be routed onward to the appropriate location as specified when using the SIP protocol, or it could be dispatched to another Application Server instance hosting an application included as part of the user's profile. The action taken by the S-CSCF is dependent on the information provided by the IMS network and its user profile information.

SIP Servlet technology provides an excellent open-standards platform for taking advantage of the adoption of the SIP protocol by IMS and its requirements for dynamic service deployment. The following diagrams are taken from the 3GPP specification titled "IP Multimedia (IM) Session Handling; IM Call Model" [1] and represent the main roles an application can assume when deployed for service in an IMS network.

Figure 6.2 illustrates an example in which the S-CSCF receives a SIP request on behalf of a user. The request is then sent to the Application Server using the previously described ISC interface. In this particular example, the Application Server is acting in the role of a terminating SIP entity. From a SIP signaling perspective, this role is known as a User Agent Server (UAS). You might remember the introduction to the role of User Agent Server in Chapter 3 and how a SIP Servlet-based application can act in this specific role. This role results in the application's generating the final SIP protocol response and sending it upstream. The SIP request is not sent any farther downstream.

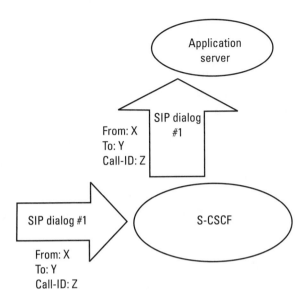

Figure 6.2 IMS terminating [1].

Figure 6.3 illustrates another role that was discussed in Chapter 3. In this case, there is no incoming SIP signaling from the network to the S-CSCF, and the IMS application deployed on the SIP Application Server is responsible for generating the initial SIP request. The trigger for generating the SIP request can be anything from the result of a HTTP converged application to a JEE JavaBean

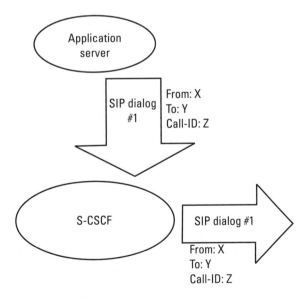

Figure 6.3 IMS originating [29].

(EJB) call to a Timer firing. (This role is known as a User Agent client UAC.) The third and final primary role defined by IMS is that of SIP Proxy.

Figure 6.4 illustrates the role of a SIP proxy server in the IMS architecture. A SIP request is received by the S-CSCF and is then passed to the Application Server for further processing. Once the application has completed its processing, it proxies the request onward to the S-CSCF, which then determines further downstream processing of the SIP request. Chapter 3 discusses how a SIP Servlet application using the appropriate API calls can act in the role of a Proxy. There is a fourth role covered in the IMS architecture that is also an important part of the SIP Servlet API. The role of a B2BUA plays a vital role in IMS and can be viewed as a concatenation of the roles depicted in Figures 6.2 and 6.3.

The suitability of the SIP Servlet API to fulfill the role of Application Server in IMS is quite evident when you map the previous constructs (UAC, UAS, and Proxy) defined in IMS to those that are exposed by the SIP Servlet API. The flexibility of standardized API and life-cycle management provides supreme confidence that applications not only will be developed appropriately but also will be portable across compliant implementations. There are also specific API calls that are included in the SIP Servlet API that allow for ease of integration into an IMS environment. For example, a large number of IMS implementations use the topmost SIP "Route" header to carry the name of an application that should be invoked on an Application Server. The S-CSCF might push a SIP "Route" header that looks like this:

```
Route: <sip:app1@application_server.com;lr>
```

This is one of many examples in which the application name has been included as part of the SIP URI in the "Route" header. The application name, "app1," appears in the user part of the SIP URI (before "@" symbol). Not only could the Application Router that is part of the receiving SIP Servlet container (as introduced in Chapter 4) select an appropriate application based on this information, but the SIP Servlet API also provides the "SipServletRequest.getInitial PoppedRoute" method call to allow applications to gain such important contextual knowledge.

In Chapter 4, "Application Router," we discussed the concept of routing regions, which enable applications to be invoked based on the originating and terminating status of the user. This concept also aligns perfectly with the IMS architecture, which bases application composition on servicing users in the same states.

The companies developing the IMS framework have a lot of input into the development of SIP Servlet technology. While SIP Servlets is not wholly dependent on IMS, it certainly includes appropriate levels of support for the architecture.

Finally, a logical entity known as the Service Capability Interaction Manager (SCIM) has existed in IMS for a number of years. While it has never been fully

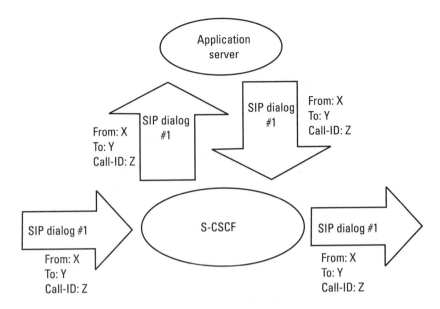

Figure 6.4 IMS proxy [29].

defined, its role has been greatly discussed in various industry forums, most recently resulting in investigation work being carried out in 3GPP.

Figure 6.5 clearly illustrates the role that was intended for a SCIM: It was intended to be a stand-alone entity that is deployed on the ISC interface between the S-CSCF and an Application Server. SCIM was intended to provide service coordination when a number of Application Servers are being used, to provide a single complete service to the user. The multiple Application Servers making up a single service are then hidden from the rest of the network by SCIM, which acts as an integration point. Recently, 3GPP has done more investigation, documented in TR 23.810 [2], into the role of Service Brokering (and SCIM) in its architecture. While no clear standardization effort has defined a SCIM, the SIP Servlet API provides an appropriate tool kit for creating service brokering functionality. The introduction of the Application Router role (as discussed in Chapter 4) provides a blank canvas for specifying service brokering capabilities.

The dedicated Application Router API also has tools that enable the sequencing and coordination of applications using SIP that are potentially hosted on external servers. For example, as introduced in Chapter 4, the "SipApplicationRouter.get NextApplication" method call returns an instance of the "SipApplicationRouter Info" class. As part of this returned information, the previously discussed "Sip ApplicationRouterInfo.getRouteModifier" method returns context to the SIP "Route" headers that are to be pushed into a SIP request (which can be obtained using the "SipApplicationRouterInfo.getRoutes" method). The value of "ROUTE_

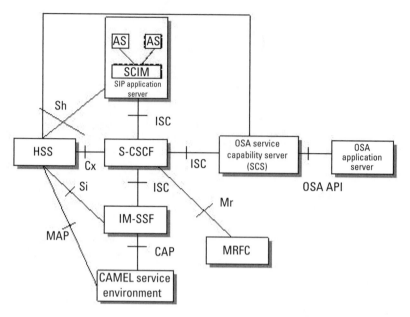

Figure 6.5 IMS SCIM.

BACK" instructs the container not only to push the specified SIP "Route" headers but also, as the last entry, to push a SIP header that will result in the SIP request's returning back to the container. A container can push as much state into this returning SIP "Route" header as required so that a sequencing chain can be resumed at a later time. This improves efficiency in the container, because state does not have to be tracked. This allows an Application Router not only to dispatch requests to applications that reside within the SIP Servlet container instance but also to include as part of a sequencing chain applications hosted on independent servers. This powerful utility, along with the rest of the SIP Servlet API, enables appropriate service brokering (SCIM-like) functionality to be developed. What is also certain is that future innovations and requirements in this area will be catered to by the technology.

References

[1]   TS 23.218, IP Multimedia (IM) Session Handling; IM Call Model; Stage 2, 3GPP.

[2]   TR 23.810, Study on Architecture Impacts on Service Brokering, 3GPP.

7

SailFin 101

The SIP container used throughout the book is the SailFin open source SIP and JEE 5 bundle. All the relevant source code is available for download for the curious reader. The server is based on the Sun Glassfish application server-code base with the addition of the SIP components that are conformant to JSR 116 and JSR 289 standards.

To find the code for a compilation or a binary bundle, you can go to https://sailfin.dev.java.net and download it from there. There are various other resources, such as "how-tos," samples, report issues, mailing lists, and architectural documents in wikis that can be of interest to developers that like to delve a little deeper. We recommend reading the next chapter on SailFin architecture to understand how a container works behind the scenes. This will give you a better understanding of the architecture and help you find more information relating to the SailFin server and its associated community.

A binary version is recommended for running the samples in this book, but if you'd like to build one from the source, that will work as well. The samples have been verified with version "b53"; any newer version should also be (backward) compatible.

If you prefer to use the source code, then check out the latest (or latest milestone build from a stable branch): https://sailfin.dev.java.net/Build_Instructions_for_SailFin.html. You can also download a promoted build from the list (see Figure 7.1). It should be pretty safe to pick the latest, since the promoted builds have been run with an extensive set of tests to prove their quality. There are more than 400 tests that are executed, and if any major flaw is uncovered by the build system, the system then prevents that build from being promoted.

To download a ".jar" file, you can follow the instructions from the site, which look something like this:

```
java -Xmx256m -jar sailfin-installer-v1-b53-darwin.jar
```

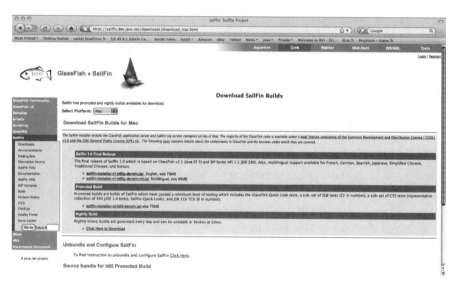

Figure 7.1 SailFin community download page.

(Don't worry: This is the installer, which consumes that much memory only while unpacking.)

- Now you should read the license and, in order to run the samples, accept (see Figure 7.2).
- Scroll down, and push the accept button.
- Now jump into the SailFin directory.
- The next step is to use "ant" to setup your server domain.
- If you already have Apache Ant Version 1.6.5 installed, you can skip this step.

Otherwise, because Ant is bundled with the binary, first you must make sure it is "runnable."

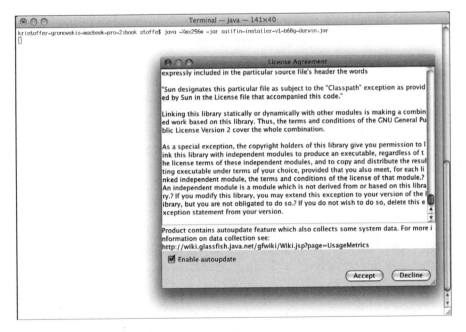

Figure 7.2   License prompt dialog when installing.

Depending on the Operating System being used, a command could look like this one, from a UNIX-based system:

```
chmod -R +x lib/ant/bin
```

Now it is time to configure a domain, and for this book we will focus on the development profile in which there is only one instance of the application server. The SailFin container also supports multiple domains, but it is beyond the scope of this book to provide such examples. If you retrieved the source code from the Concurrent Versions System (CVS) and built the system, the last step creates the domain for you, so don't execute this step but instead continue with starting the server.

In the case of a CVS image, the SailFin install directory is under "publish/glassfish."

So now let's simply create a default domain:

```
lib\ant\bin\ant -f setup.xml
```

Note This is the same as calling the default "all" target.

A directory, called "domain," is created where all your applications will be deployed and where the container is configured from and stores all the logging. Another thing that the "ant" target is doing is populating the "bin" directory with various useful scripts and binaries. The most important one is the "asadmin" command! This is the Command Line Interface (CLI) interface for the application server and the primary point for starting it.

The server is started by the following command:

```
bin\asadmin start-domain
```

Note If you have more then one domain, then you have to let the server know which domain. The default one is called "domain1," and the command is equivalent to "asadmin" start-domain domain1.

Now you can follow the startup process by tailing the log:

```
domain/domain1/logs/server.log
```

Note If you are running Windows, then no standard application will be provided, but there are some nice utilities like the "BareTail" that do this.

The log should not contain any ERROR or FATAL level of messages, and the last line should state that the server is ready to handle traffic.

The next step is to verify that the standard SIP sockets 5060 udp, 5060 tcp, and 5061 tcp Transport Layer Security (TLS) are ready to process traffic. Issue the following command:

```
netstat -an | grep 506
```

The returned list should look like this:

```
wdhcp-158-69:~/Development/book/sailfin/bin stoffe$ netstat -an|grep 506
tcp46  0  0 *.5061     *.*      LISTEN
tcp46  0  0 *.5060     *.*      LISTEN
udp46  0  0 *.5060     *.*
wdhcp-158-69:~/Development/book/sailfin/bin stoffe$
```

1 If you have multiple network interfaces, there could be a potential clash. In this case, please read about how to configure network interfaces in the in-depth chapter on SailFin (Section 8.4).

Now the server is up and running, but there are no applications being deployed. There are three ways to deploy applications. By far the easiest way is the "autodeploy" function, in which the deployer simply copies a SIP Archive (.sar) file to a specific directory.

Here we are going to use the name "example1.sar":

```
cp <path>/example1.sar <sailfin dir>/domains/domain1/autodeploy
```

Inspect the log file: It should indicate that the application has been successfully deployed.

If you want to update the "example1.sar" just simply copy it over the old file. The deployer scanner will notice the changed timestamp and redeploy it. In the same manner, you can delete the file and the scanner will undeploy it.

The other alternative to deploy a ".sar" file is to use the "asadmin" command:

```
bin/asadmin deploy <path>/example1.sar
```

Redeployment uses the same command, while for undeployment use the "undeploy" command like this:

```
bin/asadmin undeploy example1
```

Note that for undeployment only the file name is used. The path and the ".sar" extension are not used. The "asadmin" command is probably the most professional approach, since everything is navigated from it in a clustered system. It would also allow you to deploy to a cluster and it is easily scriptable.

The last alternative to deploy an application is to use the "admin" application. It is a simple ".war" file executing on port 4848 on the server.

Note You might want to restrict firewall access if putting up a system so that no one can access and administrate the server from the Internet.

`http://localhost:4848/`   (see Figure 7.3)

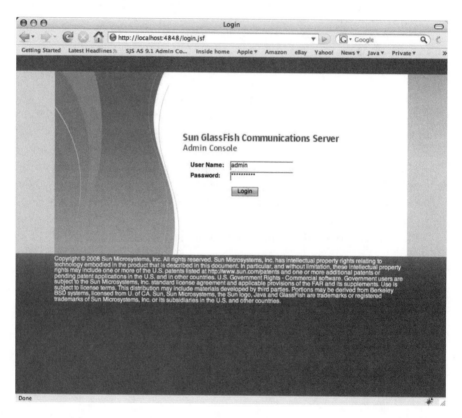

Figure 7.3 Login.

The default user name is "admin," and the password "adminadmin."

It can be quite nice to use a graphical tool like this to deploy and undeploy, since it also provides graphical feedback on the success of the deployment. Another benefit is that there is a list of other deployed applications that can be visualized. It is also useful when deploying applications from another host: Simply open a browser on the host where the ".sar" file is located, and instead of the local host URL, enter the IP of the host that runs the SailFin server (see Figure 7.4).

OK, so now there is an application server running and an application deployed. Another interesting thing to do is to set the log level.

There are various logs that can be enabled, but for doing SIP Servlet development, there is one that is most important:

SIP

The log is using the standard "java.util.logging" framework from Java SE, and by default the log level is INFO. If you want to trace SIP signaling, the recommended level is FINE. For most logging, the FINEST level can be set, but then various cleanup threads will dump out their output, so FINEST should be used only in extreme situations.

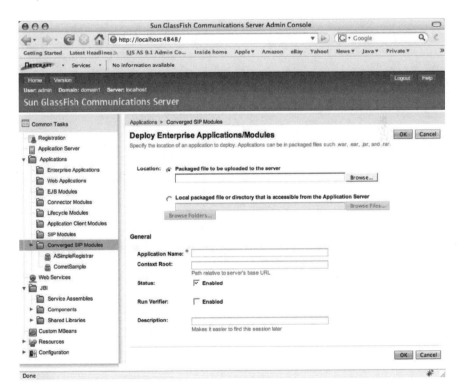

Figure 7.4 Deploy a SAR.

There are two ways to toggle between the different levels: One way is to use the "admin" Web application to change the level (see Figure 7.5).

In the same way, there is also another logger that might be interesting for a SIP Servlet developer, especially if you intend to develop or even deploy a third-party Application Router: the Application Router "extended information."

Application Router

The logger for the Application Router "ar" can be altered in the same manner as the one for SIP to yield more information in the logs. Also note that the SailFin server supports dynamic updates, so both ways of changing the logging alters the level instantly: There is no need of a restart of the server. In the same way, it is recommended not to leave a production system running on any of the FINE levels, since they produce significant amounts of data; the logs can fill up a partition, resulting in server crash. However, for development purposes, the more information the better.

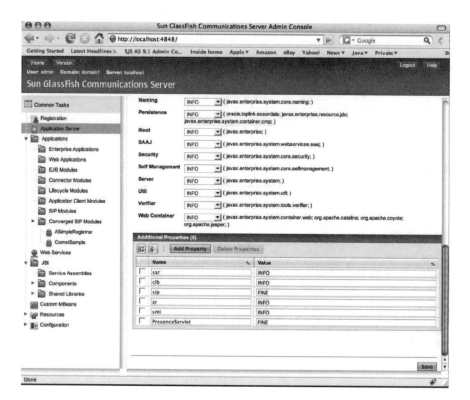

Figure 7.5 Log setting in the Admin Web Console.

It is also worth thinking about what happens if multiple applications (".sar files") are deployed in a system. Then the attention goes to the installed Application Router. SailFin comes preconfigured with a "zero config" Application Router called "alphabetical AR." You can pretty much guess what the router will do: It will probe every application with the incoming message in alphabetical manner, such that an "a.sar" file will be handed the SIP request first, then the "b.sar," the "c.sar," and so on. This is quite nice, because you don't have to read Chapter 4, "Application Router," before making your first "hello world"-style application. However, eventually, when your skills evolve, there might be a reason to move to a more complicated AR or even to write you own (for more information on writing a custom AR, look at the next chapter).

The JSR 289 standard specifies a Default Application Router (DAR) that is also available with the SailFin bundle. However, the disadvantage of this DAR is that, when an application archive is deployed, then it also has to be configured in the AR configuration file according to the SIP Servlet 1.1 specification. This is a bigger obstacle, and therefore the simpler alphabetical AR is better to enable by default.

If you don't want to worry about multiple applications, here is a pretty handy thing to do:

Make sure you followed all the steps and that the server is running fine.

Now undeploy all applications and stop the server. All the configuration changes are backed up in a file celled "domain.xml" file.

Now make a backup copy of the file:

```
domains/domain1/config/domain.xml
```

Continue to deploy applications and test around.

Whenever the configuration is broken or you want to restart from a clean sheet, then do the following:

Go back to the install directory and call the Ant setup again:

```
lib\ant\bin\ant -f setup.xml clean-runtime all
```

Copy over the backed-up "domain.xml" file.

Now start the server, and you are ready to go.

Note It will throw away your previous "domain1" folder with all the applications, logs, and configurations you had!

Cleaning up the configuration with Ant build script is a powerful way, when executing examples from this book, to ensure nothing is contaminating the execution from the previous example.

If you did some additional reading of the instructions available in SailFin, then you might have noticed an additional "asadmin" command:

```
asadmin start-database
```

This starts the built-in Derby SQL database, which is needed for any example and even for your own code that uses EJB3.

Since SailFin is a complete JEE 5 server, you can make @EJB and @Entity Manager annotations within your SIP Servlets and bundled classes in the ".sar" archives. Most examples do not need the database, but there is no harm in starting it (except for some consumed RAM and CPU, that is).

The database should open up a TCP port on 1515, and you can make sure it is operational by issuing a "netstat" command.

Other then this, there are also integration packages for NetBeans 6 (NB6) that might be of interest and that make SIP Servlet development easier. You can download NetBeans for your specific operating system from the site www.netbeans.org/.

Then in the SailFin install directory you can find additional plug-ins:

```
lib/tools/netbeans
```

Install all additional ".nbm" NetBeans module files. This can be done from the online repository or from the local files in your SailFin bundle (see Figure 7.6).

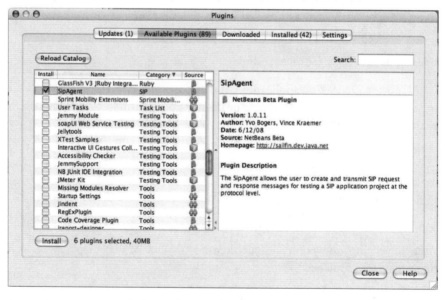

Figure 7.6 Installing SIP-related NetBeans modules.

In this package, there is SIP Servlet support for NetBeans and also a simple SIP client that can be useful for testing purposes.

Under "Server:" you can choose the SailFin install directory (see Figures 7.7 and 7.8).

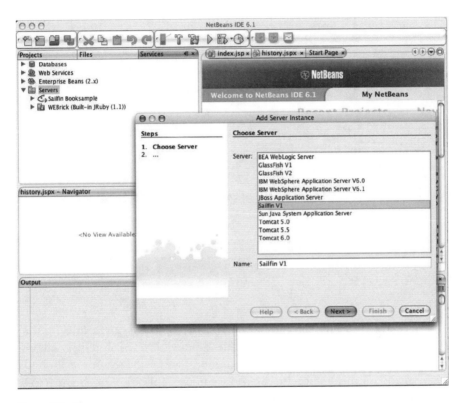

Figure 7.7 Step one.

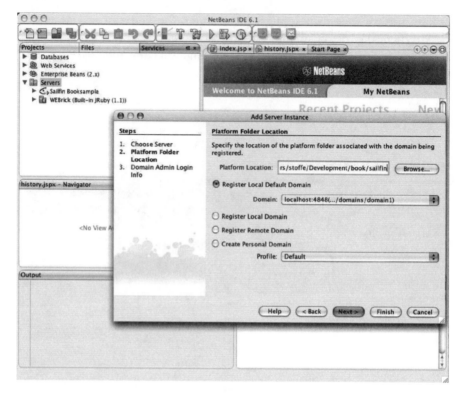

Figure 7.8 Step two.

This will enable the server to be started and stopped. It can be used for running the server in debug mode or in profiler mode. (NB6 has a pretty good profiler add-on.) Applications can be deployed/undeployed, and because it is also a full JavaEE 5 environment all of the enterprise edition application can be managed.

We talked about the EJB database support, and from NetBeans the database could also be started. (Actually, NetBeans will start it automatically for you when you add an EJB to your project. It is good to know what is happening behind the scenes; without this knowledge, it might not work in the production system, since you did not start the database yourself.) There are also tools that are useful for looking at the created tables or to determine whether there was any problem with the entity mapping.

Other than that, the primary benefit of NetBeans with the Sailfin SIP specific plug-ins is the Wizard for creating a SIP application. It will generate a project file with a ".sar" archive as the target, and it can help you with the creation of a SIP Servlet skeleton. At the same time, a "sip.xml" file will be generated together with a "web.xml" file, which can be used if the application will be an HTTP application converged with HTTP Servlets.

Note   Actually, the current plug-in creates the "sip.xml" and "web.xml," but the archive ends with a ".war." It is still possible to deploy as a ".sar," but if you want to change it, you have to go to the properties manually (see Figure 7.9).

The ".war" file is in two places—"war.name" and a few lines down in "war.ear.name," both of which need to be edited and replaced with ".sar ..." instead.

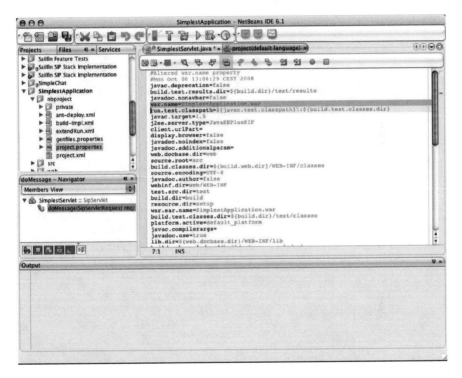

Figure 7.9   Changing the project properties from .war to .sar file extension.

Figure 7.10 is an example of what a simple application that sends a 200 OK for an incoming SIP message would look like.

Now you just compile and deploy.

Now there is an application running.

The next thing is to have some SIP message sent on the wire. Note that your SIP application has an icon available for you. This one is the SIP test client.

For testing purposes, you can create a SIP message and send it to the running SailFin server.

If you did everything right, then the 200 OK for your message should be received by the test client.

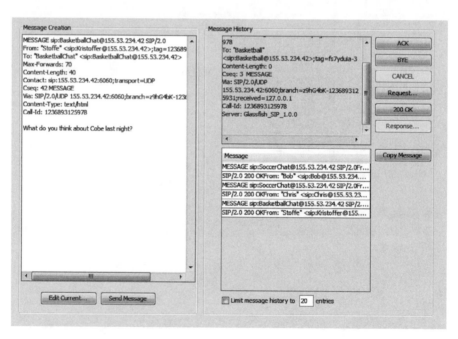

Figure 7.10   NetBeans SIP Test Agent GUI.

The simplest possible Servlet that takes a SIP MESSAGE can be created in just a few steps.

- Create the application named SimplestApplication (see Figure 7.11).

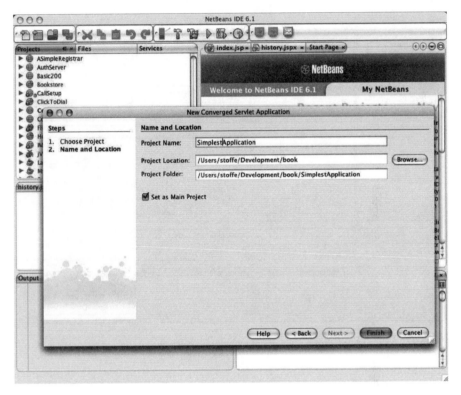

Figure 7.11 Wizard creating a SIP Servlet Application.

- Then create a Servlet (see Figure 7.12).

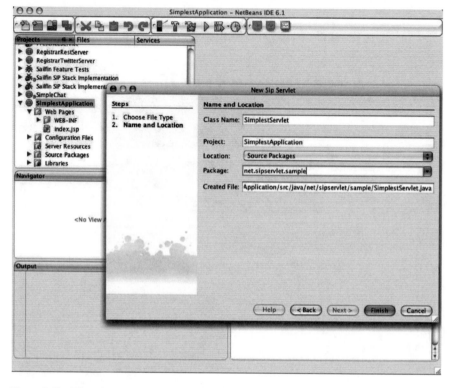

Figure 7.12 Wizard creating a SIP Servlet within the application.

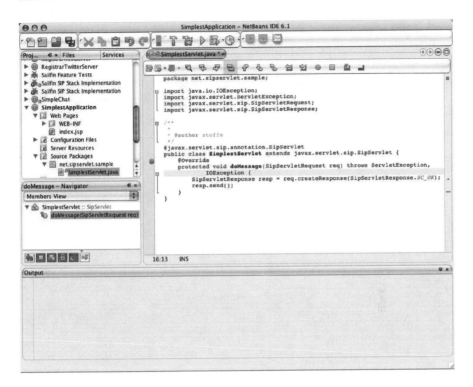

Figure 7.13 Back in editor mode filling the gaps in the generated code.

- Then fill out the skeleton (see Figure 7.13).

Since it is annotated, you don't need to worry about the "sip.xml" file.

Note The NetBeans environment can really mess up your "domain.xml" file and all the associated configuration files, so you should really think about saving away a working set. Also, it is good to start, stop, deploy, and carry out administration tasks using the "asadmin" command or using NetBeans. Do not mix, because this also can create some incompatibilities. NetBeans is a really useful tool when writing a simple SIP Servlet application. When your system becomes commercial grade, it is common to have "ant" scripts building the ".sar" files, and the deployment process for production systems is often more strict than just having the development integrated development environment (IDE) hooked to it.

These are the basic steps to get familiar with the SailFin Application Server and be up and running. You will be able to deploy the examples in this book, but we recommend having only one deployed at any specific time so not to cause interferences. Feel free to play around and create your own SIP Servlet applications.

If you want to know more about the SailFin server, you can get into more detail in the next chapter, where the architecture is revealed in order to better show how a SIP Servlet application server works and what it is really useful to know when writing production-grade systems. Remember that it is not only your application that matters but the sum of the collaboration between it and the SIP Servlet container you are running. Both have to perform in tandem to have a commercial-grade result.

8

SailFin Understanding

The inclusion of information relating to the SailFin SIP container is not in any way an attempt to promote a specific technology for realizing the SIP Servlet specification. It is primarily included because SailFin is an open source project that provides a good showcase for exactly what is required. Since it is an open source project, everything is visible, and there are no smoke and mirrors obscuring some ".jar" file. Any reader can download the full source code and see how any part of the JSR 289 specification is implemented. There is also a big difference between the developer-friendly SailFin container and the commercial product from Sun Microsystems called Sun Java GlassFish Communications Server. The source code is still open, and the commercial Communications Server uses the exact same code base. The main difference between the two products is pricing and performance. As there are no free lunches in the real world, this is also true in the software world. When used for the purpose of education and creating proof of concepts, demos, and small-scale applications, then SailFin is the perfect choice. When doing professional services, then it's all about the price and what services you want to buy. This chapter is intended to highlight the more professional aspects of a SIP Servlet container, illustrating SailFin functionality. It should be possible to do similar things with a competing commercial container, and if not, then you have a strong case in asking what is missing. Standards are great, but everything cannot be standardized in time, and understanding the implementation under the hood is essential when striving for performance, and especially in cluster deployments.

8.1 History

SailFin originated from an Ericsson IMS SIP application Servlet container built for hosting the company's multimedia applications. The key concept was to build a modular system that would scale well in a clustered environment. For that purpose, SailFin had two main design goals. The first was an optimized SIP implementation specifically built to fit SIP Servlets and not to reuse a more traditional SIP stack approach. The second objective was the scaling that led to hash-based distribution in clusters and a small memory footprint for the session objects. We will dig down much more into the architecture, but this might be handy to keep in mind if you ever download the actual source code of the container and are thinking about how this product evolved.

Initially, the SIP Servlet container did not include any JEE components other than a standard Tomcat HTTP Servlet container. Using this combination, Ericsson had deployed commercial-presence and group-list-management applications and "push-to-talk" (PTT) services based on the code. Other applications have also been developed and deployed with the Service Development Studio (SDS) (www.ericsson.com/mobilityworld/sub/open/technologies/ims_poc/tools/sds_40). The SDS is an Eclipse-based environment with IMS emulation and the SIP Servlet container bundled with some developer tools. The fact that Ericsson is not a big player in the field of developer communities, and the fact that SailFin was only really used to develop internal applications, led to partnership with a Java Enterprise Edition (JEE) industry expert. JEE technology provides a lot of useful services that can benefit a lot of coexisting in the same Java Virtual Machine (JVM) as the Sip Servlet container is executing in and vice versa.

At the JavaOne conference in 2006, there were initial discussions between Ericsson and Sun Microsystems on how such collaboration could take place. During the next year, there were several "proof of concept" projects, and finally, at the JavaOne 2007, the project was announced, and all the code could be retrieved by anyone from the open source CVS repository. Since that point, it has taken more than a year to finalize a version that is JSR 289 SIP Servlet 1.1 compliant. At the same time, the SIP Servlet 1.1 standard was not finalized, and according to the JCP process, one has to wait for the Reference Implementation (RI) & Test Compatibility Kit (TCK) before it is OK even to ship the product. So, these things go hand in hand, and one has to keep a close eye on the standardization situation, since it could pose IPR problems when property is released by the standard organization. This is one concept a freeware user has to think about, but if you buy a commercial product, then you don't have to anymore. If you have a limited budget or are doing a hobby project, don't worry: There is no point in chasing after you if you don't have a lot of money. SailFin was intended to provide a best-in-class JSR 289 container just like any of the competition but, with the development in open source, allowing people who want to have more control and, above all, freedom, a free reign to innovate. It is also quite a nice way to commoditize a market to create an appropriate pricing model. A SIP Servlet container is not as important as the actual service that one can make an actual product from but allows for a common base that provides consistency within the industry.

Look on the actual architecture and what else there is that can be useful beyond the SIP Servlet 1.1 specification.

8.2 Architecture

SIP code donated by Ericsson has been incorporated into the Sun GlassFish application server. Areas like the bootstrapping, application deployment, logging, and other container facilities are utilized from GlassFish. The SIP Servlet container has a special relationship with the Web Servlet container. This relationship evolved from the JSR 116, SIP Servlet 1.0 specification and the pure fact that both are built on top of a Servlet container implementing the HTTP and SIP protocol. The Web container in GlassFish is based on the same initial code as Tomcat, but Sun has made some significant performance and scalability improvements. It also makes use of the popular nonblocking Input/Output sockets (IO) (NIO sockets) framework called Grizzly (http://grizzly.dev.java.net). The Grizzly framework provides TCP and TLS support as well as a thread pool and a buffer pool facility, which are among its most important features. For project SailFin, this framework has been incorporated in the lowest level of the SIP stack in order to reuse the same benefits and align the source code. For this reason, Grizzly was extended with UDP support, since datagram sockets are one of the main requirements for SIP protocol support. Grizzly also provides some plugability when it comes to implementing new filters or reusing existing ones as well as exchanging pooling techniques where one example could be to deploy a real-time–capable thread pool in order to get more scheduling control.

Note By using Grizzly, it is also possible to chain all supported protocols and have only a single TCP and UDP socket visible externally. This is called port unification and in the Web deployment world can be a simpler way for only exposing a single port through a firewall. This approach has some drawbacks related to performance, since all the supported protocols need to start parsing at a single entry point, which has a penalty. For SIP there are good use cases in which this type of mechanism is an excellent idea. One such application is to add STUN (RFC3489/bis2) capabilities. STUN can be used to maintain a port mapping through a Network Address Translator (NAT) and firewall so it does not time out and cause communication problems. STUN is a simple request/response protocol over UDP that should be sent on the same IP connection as the SIP messages. In this way, a Grizzly-based STUN protocol filter could be hooked up just before the SIP parsing. This allows NAT traversal support without having to modify any of the SIP code.

When it comes to the start-up sequence of the SailFin SIP Servlet server, first a core service is bootstrapped. Then the "LifecycleListener" is invoked, bootstrapping the actual SIP container component (https://glassfish.dev.java.net/ javaee5/docs/DG/beamc.html). Once this occurs, the SIP Servlet container registers for container events (e.g., the deployment for new SIP Servlet applications). The SIP Servlet container also needs to indicate the support for new types of application archives. The SIP Servlet archive, as it is defined by the SIP Servlet specification, needs to be recognized. The Java class files and libraries included in it are loaded appropriately, and the sip.xml deployment descriptor file is parsed and acted upon. The SIP Servlet container then needs to hook up with the annotation scanning framework, and if any Web components are present, they need to be delegated to the Web container so that they will be properly loaded.

Another important linking point is to set up the ServletContext for the deployed application. This is a quite powerful feature that allows all HTTP and SIP Servlets inside the same archive to access the same Servlet Context object. In the boot-up sequence, the SIP container will also tie resources like the SIP Factory and the supported methods to that actual Servlet Context as attributes, as specified in the SIP Servlet standard. One thing that has been introduced with JSR 289 is that an instance of the "SipFactory" interface and also the "SipSessionsUtil" interface needs to be registered with the Java Naming and Directory Interface (JNDI) framework so that it can also be annotated with the @Resource annotation. By connecting this together, the SIP Servlet container will now receive all the lifetime information from the application server, such as undeployment of applications, reconfiguration, and halting the system.

Another component that is part of the integration is the ability to deploy Application Routers as defined in SIP Servlet 1.1 (JSR 289). The Application Router is packaged in a ".jar" file and uses the service provider framework from the JAR specification. We will look further into how this framework is used in writing a customized Application Router later on in this chapter.

Continuing on the integration, one key aspect of the SailFin server is its configuration. It is all controlled from a file called "domain.xml" (<SailFin install dir>/ domains/domain1/conf/domain.xml). Now, the simplest case is one in which the container is used in developer mode; then there is only one application server configured. It is also possible, even if there is only one application server, to have multiple domains defined, but for SIP, this would require the container to be configured with nonconflicting listener ports for the incoming SIP traffic. The other operation mode is when the application server is deployed as a cluster. Then the SailFin configuration is more complex, but each member of the cluster always has the same applications deployed. So, in this sense, the SailFin server is a homogenous application server. (We will cover the more advanced clustered configuration in more depth later when we discuss cluster deployment.) With this in mind, SailFin is designed in such a way that it has a Domain Admininstration Server (DAS) that is the central point of administration and configuration. The DAS is responsible for maintaining the Document Object Model (DOM) tree configured by the "domain.xml" file and distributing appropriately to all nodes in the cluster. For this purpose, every physical server node (machine) has a helper process called the Node Agent to synchronize the various commands. The commands can be things like the deployment of a new application or the reconfiguration of a variable. In every machine there can be multiple instances of an application server running. If they are running on the same machine, then the same requirement applies as for running two domains: that the network listener ports and other file descriptor resources do not clash.

In a case in which the developer mode is chosen at installation time, all this is collapsed to a single application server running both the DAS application and the "real" application that the developer deploys. One of the DAS applications is the Web-based "admin" console that can be found at http://localhost:4848 (as long as the default values used are "user: admin" and "password: adminadmin").

With this application, the administrator of a system can monitor, deploy new applications, and change logging levels, as shown in Chapter 7, but also change pretty much any of the capabilities in the "domain.xml" configuration file. Internal state information can be collected based on being retrieved from the running container as MBeans. Of course, since this is a fully compliant JEE 5 server, the framework also allows for JSR 77 and JSR 88 styles of administration.

In a more professional system, things are normally more scripted and automated. For that purpose, SailFin is shipped with a utility called "asadmin." It resides under the SailFin base directory in the "/bin" subdirectory. The "asadmin" command provides a scriptable interface to the DAS server. All of the "domain.xml "configuration file information can be accessed as well for setting up a new domain and adding application servers to a cluster. The "asadmin" uses a JMX remote connection, and once a value is entered, it is dispatched to the DAS for execution. The command will return with a success or a failure response. In the case of failure, more information can be found in the logs of the application server. And this is where we will go next, to obtain a better understanding on what happens during failures.

Note The "domain.xml" is a file that is backing up the current state of the DAS configuration. The only safe time to modify the file is when the server is switched off. At all other times, there is a risk that a change will be ignored and overwritten, and in the worst case, the entire file can be corrupted. The preferred way of modifying the information is to use either the "asadmin" command or the admin Web Graphical User Interface (GUI).

8.3 Logging

The logging in the SailFin application server facilitates the standard Java "java.util.logging" package from the JVM. The default logging level is set to "Level.INFO." This has an impact, since when making logging calls from the developed applications, it is recommended to use by best practice advice the same framework so that all the log messages end up in the same log file. Having them delivered into a single log makes it much easier to trace errors and follow the flow of the application being developed. It is also worth mentioning that the "Servlet.log()" method will log at a sufficient level for the printed string to turn up in the log. If the "Servlet.log()" would use the Level.FINE level internally in its implementation then the printed string would not be written to the file by default. The servlet developer would be confused as to where it ended up, and this is a tricky thing with logging that the logging framework does not append printouts depending on what is its current log level. As long as the log level is set to Level.INFO everything that the Servlet logs gets, gets appended to the log file.

Note   When logging, it is very important to have a good strategy for making a successful deployment for a production system. It is also worth considering the level that should be designated the default in a production system and the fact that doing large amounts of logging can consume a massive amount of CPU and disk space and even, in the worst case, grow out of control, bringing the entire machine down. Configuring a rotating log and limiting the disk storage is a good strategy. Also, dividing the log so that serious errors are logged in a severe enough level to be noticed, but then the actual stack pumps are on a FINE level, only to be produced when there is someone analyzing and scoping down for a particular error, is probably a good idea.

When it comes to SailFin, the logging can be configured at run time. The application server supports dynamic configuration updates. This feature can also be benefited by the application developer; however, note that not all configuration variables will be noticed and could need a server restart before they come into effect. When it comes to logging levels, they are dynamically updated and obeyed. So the two methods recommended in Chapter 7 are to use the "asadmin" command and, alternatively, the Web admin GUI. (The third alternative is to stop the server, manually edit the "domain.xml" file, and then restart the server.)

The following is an example illustrating how the standard SIP logging level is changed from the default value of "INFO" to print the "finest" information (i.e., at the most detailed level available).

```
> asadmin set "server.log-service.module-log-levels.property.sip"
=FINEST
```

The output should confirm the change:

```
server.log-service.module-log-levels.property.sip = FINEST
```

Now, in the same way, if you want to have an application-specific log level, it can be added like this:

In the Servlet:

```
Import java.util.logging.
Logger log = Logger.getLogger("MyAppLogger");
```

And then there is the actual "asadmin" command to enable the more detailed, application-specific logging"

```
> asadmin set "server.log-service.module-log-levels.property.MyApp
Logger"=FINE

asadmin set "server.log-service.module-log-levels
```

There are five SailFin specific logger categories: SIP Container (sip), Application Router (ar), SIP Message Inspection (smi), Converged Load Balancer (clb), and SIP Session Replication (ssr):

```
<property name="clb" value="INFO"/>
<property name="sip" value="INFO"/>
<property name="ar" value="INFO"/>
<property name="ssr" value="INFO"/>
<property name="smi" value="INFO"/>
```

The SIP Message Inspection shows an extensive trace for how a SIP message is handled by a container thread, while the SIP Session Replication (SSR) shows the synchronization activities in the cluster.

8.4 Network Configuration

Another part that is of an interest is how the network listening endpoints are configured. By default, SailFin listens on the "IP ANY" interface 0.0.0.0 and TCP and UDP port 5060, as specified in the SIP protocol. It opens up a TLS port on port 5061, also according to the standard default ports. Even if the listeners are opened up in a general way, so that a machine has multiple network interfaces, there has to be some address that is inserted into the SIP headers stating its own communication address. For that purpose, SailFin has a guessing algorithm that tries to find the most appropriate address within the available interfaces. Sometimes, as in the case of using a hypervisor like VMWare, the default-picked address is wrong, but in most cases it is sufficient.

So, let's take as an example one home server that is running as a SailFin server connected to the internet while at the same time being a DHCP host and a wireless access point to other devices. It would then have an address and IP from the local ISP—let's assume 212.145.54.56—and then also provide a home network of 10.0.0.1. The guessing algorithm will pick 212.145.54.56 and use it when it inserts information into the SIP "Via," "Contact," "Record-Route," and other such SIP headers indicating its own server address. So, in this example, a SIP INVITE message originating from a SailFin server would have a contact as follows:

```
Contact: sip:212.145.54.56:5060
```

Now, imagine that you have multiple interfaces and the wrong one is used. Another scenario is when there is a NAT or IP Sprayer Load Balancer that is the interface to the network the SailFin server is deployed on. In that case, there might be a need to insert the IP Sprayer address, or it may be even more useful to put in a DNS name so that it is easier for people to remember.

In order to overwrite the default behavior, there are actually two different mechanisms that allow a user to tweak the default configuration. SailFin has a "sip-listener" element in the "domain.xml" in which a listener by default is configured like this:

```
<sip-listener address="0.0.0.0" enabled="true" id="sip-listener-1"
port="5060" transport="udp_tcp"/>
```

Imagine that someone wants to build a SIP-firewall kind of application; it would really matter on what physical interfaces the connections are established. We can take the same home server example as before, but also remember the scenarios of multiple domains in the same server or of a machine with multiple application server (AS) instances. This configuration would also have to be modified. Here is the home server SIP-firewall example:

```
<sip-listener address="212.145.54.56 " enabled="true" id="sip-
listener-1" port="5060" transport="udp_tcp"/>
<sip-listener address="10.0.0.1" enabled="true" id="sip-listener-1"
port="5060" transport="udp_tcp"/>
```

Note It would be possible to support only TCP or UDP, but according to RFC 3261, which is the specification defining SIP 2.0, a server *must* support both UDP and TCP.

This will make sure that the connection layers will be kept as two separate IP listening points. In order to get the information what kind of socket and IP protocol a SIP message actually arrived on the standard SIP Servlet method on a "SipServletMessage," ".getLocalHost()" and "getLocalPort()" can be used to find out from within a SIP Servlet application. (It is also possible to look at "Via" headers, but it is not recommended, because the API should be the preferred way.) In the other direction, when messages are sent, the JSR 289 now allows for the multihost support that can let the Servlet programmer choose an address and in that way implement a SIP firewall.

Note The SIP Servlet 1.1 method "setOutboundInterface" should be used, but at the time of this writing, it has not yet been implemented in SailFin.

The other configuration possibility is to decide what the address and port used for identifying the SailFin server should be. The configuration element looks like this:

```
<sip-container external-sip-port="5060" external-sips-port="5061">
```

But there is another possible attribute that can be set like this:

```
<sip-container external-sip-address="sipservlet.net" external-sip-
port="9090" external-sips-port="9091">
```

A contact header from a message generated from a server would yield a header like this:

```
Contact: sip:myService@sipservlet.net:9090
```

The follow-up question: What can be done with this? One possibility is that a DNS entry can be provisioned in a way that would yield multiple server IP addresses for a lookup. More on this is described in an example later in the chapter, in which DNS and Service Record (SRV) are used in a clustered setup.

The last part of the configuration of the network listening points has to do with TLS. The TLS listener is configured with these lines in the domain.xml:

```
<sip-listener address="192.16.149.111" enabled="true" id="sip-
listener-2" port="5061" transport="tls">
  <ssl cert-nickname="slas" client-auth-enabled="false" ssl2-
enabled="false" ssl3-enabled="false" tls-enabled="true" tls-rollback-
enabled="true"/>
</sip-listener>
```

It is very similar to the configuration of a UDP and the TCP listening point. TLS is optional to support when coming to the SIP standard, but in SailFin it is enabled by default, since there is no harm in accepting connections on TLS by default on the standard port 5061. That said, if the security is taken for real, then this is not enough. The most important thing for secure communication is that a client that is connecting itself receive an identity certification from the server it is trying to establish the connection with. For example, say a company like Ericsson sets out a SIP server for all the traveling employees to use. Now, being on the road, privacy might be a nice thing to have so that open Internet users can't see where a connecting client is setting up a SIP session. TLS will help you with that, but since it is on the open Internet, someone might try to do a man-in-the-middle attack and pretend they have the ericsson.com SIP server the user is trying to access. This would destroy the entire idea of using TLS from the start, because having an encrypted socket stream with the wrong person is as bad as having a normal TCP connection with the same. For that reason a certificate identifying the SIP host would be used. In the configuration, there is this "cert-nickname" attribute that by default points to the "slas." This is a certificate that is generated by default in the installation step of the SailFin server. This certificate is used both for the SIP server and for the HTTP server so that any HTTPS browser accessing the SailFin server would receive the very same certificate ("cert"). (Of course, they can be separated by reconfiguration of the defaults.)

If you are in control of both the client and the server, then this certificate might be enough. If you do not have full control over the clients or you do not want to distribute your public certicate key and install it on every device, then it needs to be signed by a Certificate Authority (CA).

This is a normal operation in the HTTP world, and your browser will warn you if this is not the case. After you generate your private and public keys, they would be sent to a global CA (e.g., VeriSign), who will try to establish your real identity. In the previous example, Ericsson would pay a CA to prove its identity. All browsers and also the SIP stack should come with the public key of at least a couple of the global CAs. Even if the Ericsson.com server says it is serving for the domain Ericsson.com, you should not trust it unless there is a public key that matches or it is signed by a CA and that resolution proves that the CA guarantees that it really is Ericsson.com you are talking to.

The other thing that can be configured is whether clients should also authenticate themselves. Now this is not as often used, but in some cases, one would really want to be sure that the TLS identity that is trying to register and the "From" SIP header user is one and the same. For this purpose, the "client-auth-enable" attribute should be set to "true"; then, in the TLS (SSH) handshake, the server will mandate a valid client cert. In the SIP Servlet standard, the client cert would be put in a "SipServletMessage" attribute ("javax.servlet.request.X509Certificate" or "javax.servlet.response.X509Certificate") in X509 class format.

Knowing both sides of a TLS connection can be really useful. The next step that is often confused in SIP with the TLS is the SIPS protocol URI part. A "sips" URI scheme indication specifies that, if it is found in a SIP URI then the next hop to route should be handled in a secure manner. The secure (ref to RFC 3261), mechanism could be TLS, or alternatively, it could be an Internet Protocol Security (IPSec) specified connection. Since IPSec is not visible at the Java level, the SIP container cannot know whether an underlying connection is using IPSec. For that reason, the SailFin container will always attempt to set up a TLS connection when it discovers a SIPS URI as the next point at which to route. A SIPS URI could look something like this:

```
Contact: Stoffe <sips:kristoffer.gronowski@sipservlet.net>

<domain>
 <configs>
  <config dynamic-reconfiguration-enabled="true" name="server-config">
   <sip-service>
    <access-log format="%client.name% %auth-user-name% %datetime%
%request% %status% %response.length%" rotation-enabled="true"
rotation-interval-in-minutes="15" rotation-policy="time" rotation-
suffix="yyyy-MM-dd"/>
```

```
     <sip-listener address="0.0.0.0" enabled="true" id="sip-listener-
1" port="5060" transport="udp_tcp"/>
     <sip-listener address="0.0.0.0" enabled="true" id="sip-listener-
2" port="5061" transport="tls">
        <ssl cert-nickname="s1as" client-auth-enabled="false" ssl2-
enabled="false" ssl3-enabled="false" tls-enabled="true" tls-rollback-
enabled="true"/>
        </sip-listener>
        <sip-protocol default-tcp-transport="false" error-response-
enabled="false">
        <sip-link connection-alive-timeout-in-seconds="120" max-queue-
length="50" write-timeout-in-millis="10" write-timeout-retries="25"/>
        <sip-timers t1-in-millis="500" t2-in-millis="4000" t4-in-
millis="5000"/>
        </sip-protocol>
        <property name="accesslog" value="${com.sun.aas.instanceRoot}
/logs/sipaccess"/>
     </sip-service>

        <sip-container>
     <session-config>
      <session-manager>
       <manager-properties/>
       <store-properties/>
      </session-manager>
      <session-properties/>
     </session-config>
     </sip-container>
    </config>
```

8.5   SIP Container Architecture

Since this is a book related to programming, you as the reader know that configuration possibilities are always limiting. Since we are dealing with an open
source SIP container, we have the possibility to exploit all of its features. For that
purpose, we will be taking a closer look at the architecture. Most containers have
some hook mechanisms, but generally they are not all that flexible. The SailFin
SIP container is built in a layered architecture design pattern following the interceptor pattern principle. Since the interceptor was developed entirely for the purpose of being in a SIP container as opposed to a normal SIP stack structure, the
common format for it is the "SipServletRequest" and the "SipServletResponse"
object class. The basic idea is that layers are stacked together, forming a complete
SIP stack and, at the same time, a JSR 289 SIP Servlet container.

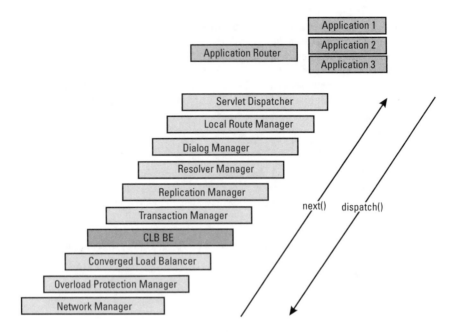

Figure 8.1   Layer Architecture inside the Sailfin Sip Servlet Container.

Figure 8.1 is a depiction of the default layers providing the complete container.

The entry point is the Network Manager layer. It parses and frames each incoming SIP message on the network listening points previously specified until a complete message is read. For this purpose the Grizzly, NIO layer is used by default together with a thread pool (com.ericsson.ssa.container.NetworkManager).

Note   There are two actual Network Manager implementations available: "com.ericsson. ssa.container.GrizzlyNetworkManager" and ".ericsson.ssa.container.OldNetworkManager" which is the legacy implementation.

When one message is framed completely by the parser (com.ericsson.ssa.sip.SipParser), it is relayed to the next layer by the Grizzly filter (com.ericsson.ssa.MessageProcessorFilter).

Make a note that the Network Manager is only the first layer, so the resulting "SipServletRequest" or "SipServletResponse" is not a complete object that would satisfy the SIP or the SIP Servlet specification. There are several fields that would still yield null values, for example, an application requesting a related "SipSession" object. Among other duties, the Network Manager sets the local and remote endpoints for communication according to the JSR 289 specification. The Network Manager also sets the TLS client cert if it was requested, and it makes sure to keep a reference about what IP stream a request arrived on so that a response can be sent in the opposite direction.

Network Manager also implements parts of the IETF draft on outbound proxy for connection reuse to deal with NATs and slow-establishing TLS client sockets.

We will come back to the Converged Load Balancer layer in a bit, since it is optional and not taking an active part in a developer single-node-instance configuration. The next mandatory layer is the Transaction Manager. Its responsibility is to keep the SIP transaction state machine specified in RFC 3261 and handle retransmissions so that messages do not reach the Transaction User (as specified in RFC 3261 [1]) layer (i.e., that two Servlet instances do not act on a resend of the very same INVITE request). It keeps track of both the server transaction and client transaction and timers related to them. In some cases when there is no answer, it would generate a 408 response to unanswered requests after 64 T1 (default 32 seconds), as specified in RFC 3261 [1]. This is a guarding mechanism in SIP, but it is an important way of communicating for the SIP Servlet as well. In a case in which a SIP Servlet is acting as a UAC, it would originate an INVITE request, and for some reason the other side might have crashed. The container would then generate a 408 internally traversing all the Servlets in its path, and the Servlet developer could clean up any consumed resources. So, this layer is mandatory and important for consistency.

The Replication Manager layer does pretty much as its name implies. If SIP Session replication is enabled for high-availability purposes, then SIP dialog-related data is propagated in the cluster. For this purpose, a replication framework similar to that dealing with HTTP Session replication has been designed to take care of the SIP counterparts. It is based on Shoal/JXTA, and it serializes the data objects. The cluster is formed in a ring topology in which every node replicates all of its session data to its buddy (neighbor). Then, in failure events or when one node is taken out of service, the new node handling the established SIP dialogs is able to retrieve the dialog from the backup copy. SIP Servlet Timers are also stored and replicated and require special attention, since someone needs to fire them when the owning node exits the cluster.

The next layer is the Resolver Manager. It primarily acts on messages leaving rather than entering the SIP Servlet container. When it comes to forming the path into the container, the Resolver Manager examines the SIP "Via" header that the communicating client has set and verifies whether it matches the credentials received earlier by the Network Manager. In the SIP specification, the received parameter and the rport "Via" parameter is set on the topmost "Via" header (if the "rport" parameter is supported).

The Dialog Manager layer checks for the existence of a SIP "To" header tag parameter. This indicates that this message is a part of an ongoing SIP dialog that is created for SIP requests like INVITE, SUBSCRIBE, and REFER. Also, note that other SIP messages like a MESSAGE can be sent in an established SIP dialog. When a SIP dialog already exists, the same SIP Servlets that were traversed in the initial composition should also be traversed for the subsequent flow. If the dialog manager cannot find any previous information, it will return a SIP 481 session/dialog "does not exist" error response. If it is found, the Dialog Manager needs to find the direction. When an initial call coming from A to B generates SIP Servlet invocation S1,S2,S3, any subsequent signaling from A should follow that order. But when subsequent signaling, such as the SIP BYE message, coming from B is an established dialog, the Dialog Manager needs to ensure that the reversed chain is executed. In this case, S3,S2,S1 would be the result. The container achieves this by pushing the correct instances on the dispatcher stack, and then the S1,S2,S3 are executed and popped from the stack.

The last layer completing the required stack is the Application Dispatcher layer. When the message reaches this layer, it is complete, taking into account the SIP transactions, SIP dialog objects, and other aspects mandated by SIP. There are two possible outcomes when coming to the Application Dispatcher. If it is an initial request, then the Application Router would be consulted to decide which application should receive the request next. For subsequent calls, this layer is not reached and is taken care of by the Dialog Manager described previously. So the AR is in charge of receiving the request and decides that it should be handled by application A. Now the Application Dispatcher starts the layer chain in the other direction by calling the dispatch method. It locates the Servlet Dispatcher that represents application A and calls its dispatch method. The Servlet Dispatcher is consulted, as discussed in Section 2.1.3. It would either use the SIP Servlet mapping rules to find a Servlet match, or it would relay to the main Servlet for that application. The Servlet Dispatcher finds the correct Servlet instance and calls its service method. For simplicity, we could imagine that the SIP Servlet is a simple UAS with two lines returning just a 200 OK message. The other alternatives are the Proxy or a B2B, but for now we would just follow the response back.

```
Public void doMessage(SipServletRequest req) {
        SipServletResponse resp = req.createResponse(200);
        resp.send();
}
```

Behind the scenes, the correct fields are copied from the request to the response, and objects are linked in order to be able to find each other at a later point in time. After completing the linking, the "send" method will eventually start the trip down the layers in the reversed order. On the way up the layer architecture, each layer can decide to push a Dispatcher interface class. Often it is the Layer class itself, but in some circumstances it can be a delegated class, or a layer can even ignore being part of the message going out. One example in which a layer is not the one that handles the outgoing message itself is that of the Transaction Manager. The Transaction Manager pushes a server transaction that implements the Dispatcher interface on every incoming message. In this case, there is no need to search for matching transaction once the 200 OK message is being sent; just pop it from the dispatcher stack. Another layer that uses this technique is the Network Manager in the case of TCP and TLS. The same principle applies here in that if a request arrives on TCP, then according to the specification, the response should try to reuse the same socket on the reverse path. Actually the Grizzly Network Manager pushes both the Stream Response Dispatcher class that represents a connected TCP socket as well as the Network Manager. The reason for this is, if the open connection has failed, then the Network Manager will establish a new TCP connection.

The normal exit for a response is to first pop the topmost dispatcher, which is the Dialog Manager for initial requests or the ResolverManager for subsequent requests (after a visit to all the SIP Servlets by Application Dispatcher or Dialog Manager). This ensures that the SIP dialog information, "SipSession" object, and "SipApplicationSession" objects are consistent. Next in line is the Resolver Manager, which inspects the next destination to route and determines whether there is a tel URL with an ENUM phone number to translate or there is a DNS entry address to look up. In both cases, it will try until it gets a SIP URI with an IP part in the destination host field. This layer makes sure to implement the RFC 3261 and RFC 3263 lookup mechanism specified in SIP. After the Resolver Manager, the next dispatcher layer to be called is the Replication Manager. In the case of session replication, the data would be check-pointed by the Replication Manager and replicated to a buddy node in case there should be any failures of the current node.

In the case of a single node, the next layer would be the transaction layer and, as mentioned before, the actual server transaction if it is a response and, if a request, there wouldn't be any transaction yet, since it is an outgoing new request, so the Transaction Manager would be invoked and a client transaction associated. After the transaction layer, it is normally time to go out on the network, and here the Network Manager or an already established TCP or TLS connection is reused.

So this describes the core layers. You might be asking why it is useful to understand this. First of all, every SIP container has to do this kind of functionality in this order to comply with SIP RFC and the SIP Servlet standard. So, even if it's not well separated or made visible by an application server vendor toward the Servet developer, this is what happens inside a SIP Container. By understanding the architecture, one layer can be enhanced and replaced. Another socket library can be used, or a different thread pool or maybe session replication storage that uses some Structured Query Language (SQL) database instead. There is also a good chance that, if there are improvements with proven functionality, they could be donated to the community and become part of the SailFin code base.

There are two more layers in the architecture worth knowing about. The first one is the Overload Handler Manager. It tries to detect when the particular hardware the SailFin server runs on is overloaded. It uses a sampling technique to periodically monitor the CPU and memory consumption using a standard MBean interface from JEE 5. There are different thresholds, up to the first of which would be considered normal operations. When the first threshold is reached, then all initial SIP messages would be rejected (the ones that do not have any to-tag are part of an existing dialog). The idea is that every new initial request will potentially trigger a SIP dialog, making all kind of objects, which requires the server to keep consuming both CPU and memory. If the next threshold is reached, then only responses will be accepted into the server but not any SIP requests. In this step, the server has a lot to do and does not want to accept requests because they create additional transactions; however, by allowing responses through, it hopes to finalize an already existing transaction. Competing transactions would lead to the fact that the Transaction Manager removes its reference to the transaction and resources can be garbage collected. The last threshold is the critical mark where the server will throw away all requests and responses in an attempt to stay alive and to catch up on its processing tasks. This can happen for various reasons, such as that the Garbage Collection is not well tuned and the server is doing a long Garbage Collection pause. Other possible scenarios are that someone is doing a denial-of-service (DOS) attack or that your system generates massive traffic flow at the same time. This can also happen in voting applications and should be estimated for.

Now you may think that this is not the correct behavior for my application. Right! This is exactly why this chapter has been included in the book. There might be a requirement to track some other resource, since that one might be the limiting factor. It's not uncommon for there to be a database tier with a fee involved in the number of connections made or with some other counter in the database that limits the system capacity. Instead of monitoring CPU or memory, the updated version of the overload could look at the database usage.

There is another question that is important to think about. Parsing and framing a SIP message takes some time and effort, so if you decide that it should be rejected in order to protect the system, then this has to be done pretty early in the layer architecture for performance reasons. Here there is a big advantage, since making a Servlet do the same job requires the full stack to have been traversed, and then a lot of objects are created. One can ask the question, how much then do you really save by rejecting a SIP request? The next thing to figure out is what kind of response code should be propagated back. In most cases, a 408 or 503 with or without a retry after it should be used. This is one of the weak points of SIP, because the flow control is quite badly supported. Let's take a closer look at what happens:

If a destination receives a 503 SIP error response with a retry after "x" seconds, then it would wait "x" seconds before it retries that SIP URI again. In the other case, in which a 503 does not contain the SIP "Retry-After" header or send a 408 response, what would happen so that the current request rejected? It is up to the other client to retry the request, when and if it wants. Often the other side would just try again straight away. So why is this bad? If the other side is a client, then this mechanism works very well. But in a case in which the sending side is a SIP proxy server, where "x.com" sends to "y.com" and the "sip:y.com" server responds with a 503 retry after 5 seconds, the result will be that the "x.com" server has to buffer all its requests to y.com for 5 seconds. In turn, it can result in the x.com server's also getting congested and the entire network malfunctioning. It would be far better if the "y.com" server could gradually ask the "x.com" server to buffer more messages. Now it is "send me either all or nothing." There are IETF drafts addressing this issue, and a gradual percentage-based system is suggested within the working groups expanding the new standards.

This is an important concept to understand when dimensioning larger commercial systems. Even if there is now a standardized solution, a server can protect itself with an Overload Manager approach. Another thing that can be done is to use TCP-based transport, since in the TCP stack there is the sliding window. When a remote SIP server has too much to do, it will not be able to parse a SIP request from the TCP buffer fast enough, so this will result in a decreased sliding window. In our previous example, the "x.com" server will not be able to write the bytes to the socket because the underlying Operative System (OS) will tell him to wait. This will throttle the system in a way. Another good thing to keep in mind is not to send large chunks of body content within the SIP messages. For streaming purposes or to send messages with pictures, there are other more suited protocols (such as MSRP, RTP, and so forth).

Another thing that is worth noting is that the Overload Manager is also linked in on the HTTP path (org.jvnet.glassfish.comms.httplayers.HttpLayer). One good thing that could be a best practice is to also restrict the deployed applications from being accessed in overload situations.

Since a Web Servlet is competing for the same resources, like CPU and memory, it makes sense to reject them as well. There is one exception that is worth mentioning, and that is the admin application. It could be a good idea to be able to log in and administrate the system even if it's extremely overloaded. (It would probably be a better idea to use the Command Line Interface (CLI) "asadmin" to decrease a log level or undeploy a spinning application.) However, maybe your system is generating revenue. The administrator would go into conflict and consume resources, so your system would make less money. Then it would be acceptable for the admin application to be blocked too. But leave a window open if you need to access an overloaded system.

The Local Route Manager layer is a good showcase for how extending the SIP container can be made easy. It simply inspects the outgoing messages and looks at the next hop. If someone has put 127.0.0.1 or any other IP address that would end up in the very same instance of the SailFin container, then why take the penalty of serializing and deserializing? This simple layer resets the state and pushes the outgoing request back to the Application Dispatcher (AD). The AD does not care if the message was sent out on the wire or not.

The last layer that is bundled with SailFin is perhaps the most interesting one. It is called the Converged Load Balancing (CLB) layer. Use of the term *converged* comes from the fact that it is in the both SIP and HTTP paths. This layer is also one that should be intercepted as close to the parsing and framing of protocol messages as possible. What it does is act as a load balancer, distributing the incoming traffic to a specific node in the cluster. In the case of SIP, it would happen before the transaction layer, and so it is to be seen as a stateless proxy SIP server. For HTTP, this happens before the HTTP Session mechanism, so it would be behaving as a normal HTTP proxy. This layer uses a consistent hash algorithm to make sure that requests for a specific SIP URI always ends up on a particular SailFin server instance. Just to give a simple example, if a request is targeted to "sip:voicmail@sipservlet.net" and the SailFin cluster is handling the SIP "servlet.net" cluster, then it would use the user part of the URI, in this case the string "voicemail." That would be taken as input to the algorithm, and the other part is the list of available servers. The CLB layer is using Shoal to "heartbeat" the cluster, keeping a list of servers that are alive and well. From that list, for every time the function is called, it would yield server X and the nice part is that, if server X is down, then server Y will be returned. That limits the need to transfer any SIP transaction and dialog state as well as any "SipSession" or "SipApplicationSession" data (other than replication if High Availability (HA) is needed).

There are other mechanisms detailing how high availability and load balancing could be achieved, and we will talk more about it in the cluster section at the end of this chapter. For now, it is important to understand how SailFin does it, that it allows you to change the algorithm or the actual place in a SIP or HTTP message where it looks for the input string for the algorithm. In the cluster section, we will discuss what other alternatives there are, and of course, nothing prevents you from replacing this layer or using a high-end IP Sprayer hardware that already understands SIP and pretty much does the same thing. There is also the "SipLoad BalancerManagerBackEnd," which is the receiving part, so when a SIP message has already been load balanced, this makes sure that on the way out it is routed over the same front-end server that it originally came in on. This is of course so that responses arriving on TCP can reuse the same socket or, in the case of connection reuse, this would also be able to navigate to the right flow.

Now that you know what layers there are, we are going to talk about why and how to write your own layers.

8.6 Writing Your Own Interceptor Layer

We touched upon some aspects of when one would want to write his or her own layer. Here is a list:

- Replace or enhance an existing one.
- Add more logging in a place for debug purposes.
- Add a strict syntax check layer for development but remove it for production.
- Add statistics.
- Add license manager layer (e.g., my customer paid for only 10 simultaneous calls).
- A SIP client that many users have not behaving correctly when it comes to the SIP standard.
- Implement a non–JSR 289 compatible User Agent that only has one application, so it does not need AR, mapping rules, and so forth (could be to decrease footprint).
- Implement an Internet draft or a not supported by SailFin RFC.

The process involved in making your own layer requires the following steps: Create a class that extends the "Layer" interface. Then implement the methods and build a ".jar" file with the class.

This class should be put into the "<SailFin inst dir>/lib" or any other place in the Java class path. Then the "domain.xml" configuration file needs to be modified to be included in the correct location in the chain.

We start by extending the right interfaces.

Now let's look at the Layer interface:

```
next(SipServletRequest req)
next(SipServletResponse resp)
registerNext(Layer l)
```

And then we have the Dispatch interface that the Layer inherits from:

```
dispatch(SipServletRequest req)
dispatch(SipServletResponse resp)
```

Other than that, all the layers are also scanned for lifetime methods:

```
public void start()
public void stop()
```

There is also a need of making the Layer instance so that a static call to the "getInstance" method is performed in the effort to access a pointer to each layer. (This is not the nicest pattern, and one has to remember to implement this method in order for the bootstrap to be correct. This will more than likely be redesigned in next version and could use the same pattern that the AR is using.)

```
public static Layer getInstance()
```

There is also a "LayerHelper" class in the "com.ericsson.ssa.container" package, which can be used to keep track and call the next layer.

Here is the example code that pretty much performs a dummy logging layer. It will intercept each message and print out its presence to the log.

```
public class PatchupLayer implements Layer {

  Layer nextLayer = null;=

  static Logger log = Logger.getLogger("PatchupLayer");

  private static final PatchupLayer singletonInstance = new Patchup
Layer();

  // Enforce Singleton pattern
  private PatchupLayer () {
  }

  public static PatchupLayer getInstance() {
    return singletonInstance;
  }

  public void next(SipServletRequestImpl req) {
    log.severe("Do the useful code here on the way in to stack");
    req.pushTransactionDispatcher(this);
    req.pushApplicationDispatcher(this);
    LayerHelper.next(req, this, nextLayer);
  }

  public void next(SipServletResponseImpl resp) {
    log.severe("Do the useful code here on the way in to stack");
    LayerHelper.next(resp, this, nextLayer);
  }

  public void registerNext(Layer next) {
    log.severe("Register layer after the patchup.
"+next.getClass().getName());
    nextLayer = next;
  }

  public void dispatch(SipServletRequestImpl req) {
    log.severe("Do the useful code here on the way out of stack");
    Dispatcher d = req.popDispatcher();
    if( d != null ) d.dispatch(req);
  }
```

```
public void dispatch(SipServletResponseImpl resp) {
  log.severe("Do the useful code here on the way out of stack");
  Dispatcher d = resp.popDispatcher();
  if( d != null ) d.dispatch(resp);
}
```

One thing that was not mentioned before was that, in the next method, there are two calls for pushing a "TransactionDispatcher" and a "RequestDispatcher." This is in order if the layer wants to be part of the chaining going out. The reason for why there are two is that the Transaction Dispatcher is used for responses going out, while the Request Dispatcher is used for requests that are generated because of an incoming request. For example, the SIP 200 OK on an invite could consume the transaction stack, the SIP INVITE request could have first been proxied, resulting in a new INVITE SIP message leaving the container. It would then consume the request stack the dispatcher pushed previously. Eventually, the proxied destination would return a 200 OK. That one would enter the stack with the next call, but after reaching the proxy, it would dispatch the 200 OK back to the originator of the call, consuming up the transaction dispatcher stack.

Now, understanding this mechanism, one has to decide whether both cases are interesting for intercepting and whether the layer stacks should be given a pointer to this layer or to a delegating one, as in the example of the Transaction Manager handing off to the Server Transaction.

Note   The transaction stack would push a Server Transaction (ST), while for the request stack it is better to put a reference to the Transaction Manager, since creating a Client Transaction (CT) without knowing whether the application will ever be a Proxy or B2B could waste valuable resources.

When the code is complete, the following log lines will have been added to trace the progress. Note that SEVERE level is used so that there is no need to go into the configuration and to enable it in order for it to appear in the log file. There should be no other severe message,s so they should be easy to spot.

A NetBeans ".jar" project could be compiled, or if it is preferred, another Integrated Development Environment (IDE) or even an Ant "build.xml" is simple enough in this case. In this example, a "PatchupLayer.jar" is produced and copied to the library directory of SailFin.

Now it is time to go in and alter the "domin.xml," so the new layer will be instantiated. Here is what the configuration would look like:

```
<sip-container external-sip-port="5060" external-sips-port="5061">
    <session-config>
    <session-manager>
```

```
      <manager-properties/>
      <store-properties/>
    </session-manager>
    <session-properties/>
  </session-config>
  <stack-config layer-order="NetworkManager, PatchupLayer,
ConvergedLoadBalancerFactory, SipLoadBalancerManagerBackEnd,
TransactionManager, ReplicationManager, ResolverManager, DialogMan-
ager, LocalRouteManager, ApplicationDispatcher">
    <stack-layer class-name="com.ericsson.ssa.container.Network
Manager" id="NetworkManager">
      <property name="reporters" value="CallflowReporter,
SipMessageReporter"/>
    </stack-layer>
    <stack-layer class-name="net.sipservlet.sample.layer.-
PatchupLayer" id="PatchupLayer">
    </stack-layer>
    <stack-layer class-name="com.ericsson.ssa.container.Overload
ProtectionManager" id="OverloadProtectionManager">
      <property name="httpLayer" value="true"/>
    </stack-layer>
    <stack-layer class-name="org.jvnet.glassfish.comms.clb.core.
ConvergedLoadBalancerFactory" id="ConvergedLoadBalancerFactory">
      <property name="httpLayer" value="true"/>
    </stack-layer>
    <stack-layer class-
name="org.jvnet.glassfish.comms.clb.core.sip.SipLoadBalancerManager-
BackEnd" id="SipLoadBalancerManagerBackEnd">
    </stack-layer>
    <stack-layer class-name="com.ericsson.ssa.sip.transaction.-
TransactionManager" id="TransactionManager">
    </stack-layer>
    <stack-layer class-name="com.ericsson.ssa.sip.persistence.-
ReplicationManager" id="ReplicationManager">
    </stack-layer>
    <stack-layer class-name="com.ericsson.ssa.sip.dns.Resolver
Manager" id="ResolverManager">
    </stack-layer>
    <stack-layer class-name="com.ericsson.ssa.sip.DialogManager"
id="DialogManager">
      <property name="FactoryClassName"
value="com.ericsson.ssa.sip.DialogManager"/>
    </stack-layer>
    <stack-layer class-name="com.ericsson.ssa.sip.LocalRouteManager"
id="LocalRouteManager">
    </stack-layer>
    <stack-layer class-
name="com.ericsson.ssa.container.sim.ApplicationDispatcher" id="
ApplicationDispatcher">
```

```
    <property name="applicationRouterClass" value="com.ericsson.
ssa.router.AlphabeticalRouter"/>
     </stack-layer>
    </stack-config>
   </sip-container>
```

Note that, in the "stack-config layer-order" element, we have added the "PatchupLayer." The other place is a new "stack-layer" element, which also specifies the full class name with the page prefix. There is also a property set that it should be an "httpLayer." We will come back to that in a bit.

Now let's start the server and look in the log file for the lines coming into action. We can reuse the "SimplestApplication" Servlet shown in Chapter 7, which, basically, sends a SIP 200 OK response to a message. For this purpose we can use the NetBeans built-in SIP test agent. The log printout in SailFin yields this:

```
[#|SEVERE|sun-glassfish-comms-server1.0|PatchupLayer|_ThreadID=15;
_ThreadName=SipContainer-serversWorkerThread-5060-8;_RequestID=
0b03e784-d297-438d-b195-15a730c8709c;|Do the useful code here on the
way in to stack|#]
[#|SEVERE|sun-glassfish-comms-server1.0|PatchupLayer|_ThreadID=16;
_ThreadName=SipContainer-serversWorkerThread-5060-9;_RequestID=d2dea1
62-ace1-4001-9c1d-4932cc71ca77;|Do the useful code here on the way
out of stack|#]
```

Note   One interesting thing that can be noted is that the request is handled by worker thread 8, while the response, by thread 9. This is due to the asynchronicity of SIP, in which one request can yield multiple other requests and responses. For an HTTP Servlet programmer, this is a new experience.

From the log above you can see that it is intercepted both on the call-to-next and dispatch.

We are now returning back to intercepting HTTP traffic in addition to SIP traffic. The property "httpLayer" has special meaning for the container. It enables a hook to the Web container in SailFin. Other than that, properties can be added using a JavaBean style of declaration. If you look at the properties of the Network Manager, you'll see it declares a reporter's property. That one could be picked up in our "PatchupLayer" simply by adding a "getReporters()" and a "setReporters()" public function declaration.

For the HTTP bootstrapping, the "PatchupLayer" needs to implement the "HttpLayer" interface and an additional factory method,

```
public class PatchupLayer implements Layer, HttpLayer {
```

as well as the following three methods:

```
public static HttpLayer getHttpLayerInstance() {
  return singletonInstance;
}

public boolean invoke(Request request, Response response) throws
Exception {
    log.severe("Do the useful code here on the way it to http");
    return true; //Return false if not wanting to continue
}

public void onDestroy() {
    log.severe("Do the useful code here on http destroy");
}
```

Now, in order to execute the code, we can use the automatically generated JSP file from the "SimplestApplication," which simply prints "Hello World" in the browser.

```
http://localhost:8080/SimplestApplication/
```

The log result after loading the page would yield the following:

```
[#|SEVERE|sun-glassfish-comms-server1.0|PatchupLayer|_ThreadID=17;
_ThreadName=httpSSLWorkerThread-8080-1;_RequestID=0f9514d4-7a07-4a09-
aa4b-24565390e3fa;|Do the useful code here on the way it to http|#]
```

So what else could you do with this?

A more advanced responsibility for it would be to count concurrent calls. To achieve this, there would have to be a counter; then, for every ACK response for an INVITE SIP message, it would be incremented, and for every 200 OK for a BYE, decremented. It is not much of a jump to say that, if the counter reaches X, the layer starts to reject the initial SIP INVITEs—and there you have a licensing layer throttling on concurrent calls. Basically, the sky is the limit, but if you invent a really cool layer, it is polite to donate good design back to the community (unless one is making a tremendous amount of money on it).

8.6.1　Writing Custom Application Router

The JSR 289 specification allows for writing of customized Application Routers (AR). Until now, we have been using the default "AlphabeticalApplicationRouter" in SailFin. It simply chooses and sorts the deployed applications based on their names in alphabetical order. "a.sar" will be called before the "b.sar" application no matter when they were deployed or undeployed. This is quite convenient, since most architectures deploy only a single application, and they do not need to worry about any AR configuration. In a deployment with multiple ".sar" files, it might be required to write a custom AR component. This happens only when two ".sar" files are triggered for the same kinds of messages and they cannot be separated by mapping rules. If it is the same person writing the SIP applications, there is an option to put them into the same ".sar" file and to use the previously described main Servlet approach instead. If you prefer to write small, independent SIP applications that can be reusable, then it might be better to write a custom AR. We will come back to other reasons why one would want to write a custom AR after the example.

The custom AR uses the ".jar" Service Provider Interface (SPI), so we need to package our custom router accordingly, in a ".jar" file. Then we need to create a META-INF/services directory according to the service provider SPI requirements. In the services directory there needs to be a "javax.servlet.sip.ar.spi.SipApplication RouterProvider" file. This file should point to our custom "SipApplication RouterProvider."

In the file we specify the name of our factory class:

```
net.sipservlet.sample.ar.CustomRouterProvider
```

Then we need to implement the "CustomRouterProvider" interface and override the factory function.

```
public class CustomRouterProvider extends SipApplicationRouter-
Provider {
 private final CustomApplicationRouter ar = new CustomApplication-
Router();

 public CustomRouterProvider(){}

 @Override
 public SipApplicationRouter getSipApplicationRouter() {
  return ar;
 }
}
```

The provider class needs to have a public, no-argument constructor, and it needs to implement the abstract method "getSipApplicationRouter." Here we instantiate our custom AR class and return it. When this is done, then the "CustomApplicationRouter" class has to be implemented.

```
public class CustomApplicationRouter implements SipApplicationRouter {
 private Logger log = Logger.getLogger("CustomAR");
 List<String> deployed = new ArrayList<String>();
```

```
public void init() {
 log.info("Init AR called");
}

public void init(Properties prop) {
 log.info("Init AR called");
 for(Object o:prop.keySet()) {
  log.info("Property = "+o.toString()+" : value = "+prop.getProp-
erty(o.toString()));
 }
}

public void destroy() {
 log.info("Destroy AR called");
}

public void applicationDeployed(List<String> apps) {
 deployed.addAll(apps);
 for(String app:apps) {
  log.info("AR deployed app = "+app);
 }
}

public void applicationUndeployed(List<String> apps) {
 deployed.removeAll(apps);
 for(String app:apps) {
  log.info("AR undeployed app = "+app);
 }
}

public SipApplicationRouterInfo getNextApplication(SipServletRequest
req,
   SipApplicationRoutingRegion region, SipApplicationRoutingDirective
directive,
   SipTargetedRequestInfo info, Serializable stateInfo) {

 log.info("AR getNextApplication returns -> "+deployed.get(0));
 SipApplicationRouterInfo result = new
SipApplicationRouterInfo(deployed.get(0), region, null,
   new String[0], SipRouteModifier.NO_ROUTE, stateInfo);
 return result;
}
}
```

The class needs to implement the "SipApplicationRouter" interface, where the different life-cycle methods will be called on the custom router. We also define a proprietary "Logger" so that we can easily trace our AR. Then a list with the deployed applications needs to be stored. On "init" and "destroy" function calls to the AR, we simply log them. When we get a callback to "applicationDeployed" and "applicationUndeployed," we simply add the applications to our list, or remove the entries in the case of undeployment. Then we also add some logging to the deployment events to see when they are called.

The most important function is the "getNextApplication." This function gets called for every initial "SipServletRequest" object being passed to the SIP container. After returning the first application that should handle the request, if the first application did not act in the role of UAS but instead acted as a Proxy on the request (or acted as a B2BUA), then the modified request would trigger our custom AR again. For this sample, we do not inspect the request or do any advanced analysis. Instead, the first deployed application in our list would be returned.

When all the code is in place, then we need to build the ".jar" file containing our custom AR. Then we need to copy it to the SailFin "lib" directory (actually, it could be any directory on the Java path). When that is done, we should start the SailFin server:

```
bin/asadmin start-domain
```

Looking at the log "domains/domain1/log/server.log," we should now see a new entry using our "CustomAR" log level.

```
[#||INFO|sun-glassfish-comms-server1.0|CustomAR|_ThreadID=10;_Thread
Name=main;|Init AR called|#]
```

Here we can see that the container has instantiated our custom AR. When the container is up and running, we need an application to test our custom AR. For this purpose, we can reuse the "SimplestApplication" we wrote in Chapter 7.

```
bin/asadmin deploy SimplestApplication.sar
```

In the log we should find the following entry:

```
[#||INFO|sun-glassfish-comms-server1.0|CustomAR|_ThreadID=14;_Thread
Name=httpWorkerThread-4848-1;|AR deployed app = /SimplestApplication|#]
```

Now that we have deployed the application, we need to generate some SIP requests to try it out. Once more, the NetBeans SIP Test Agent can be used to generate a SIP MESSAGE request.

```
[#||INFO|sun-glassfish-comms-server1.0|CustomAR|_ThreadID=16;_Thread
Name=SipContainer-serversWorkerThread-5060-0;|AR getNextApplication
returns -> /SimplestApplication|#]
```

So the invocation works fine. If we undeploy the application, then the "applicationUndeployed" function should get called.

```
bin/asadmin undeploy SimplestApplication
```

```
[#||INFO|sun-glassfish-comms-server1.0|CustomAR|_ThreadID=17;_Thread
Name=httpWorkerThread-4848-0;|AR undeployed app = /Simplest
Application|#]
```

The last callback is the "destroy" method, and for that we need to shut down the SailFin server.

```
bin/asadmin stop-domain
```

```
[#||INFO|sun-glassfish-comms-server1.0|CustomAR|_ThreadID=18;_Thread
Name=RMI TCP Connection(13)-127.0.0.1;|Destroy AR called|#]
```

That concludes the simple sample, but certainly much more advanced application routers can be written: one such example is the "echarts.org" community, in which an entire framework for application composition is provided. The hub of the Echarts framework is an Application Router that can be programmed as a state machine.

There are other potential use cases: There could be an application that was developed independently and now needs to be run in an IMS network. The IMS specific selection can be adopted in the AR layer (like finding the IMS session case and mapping it to a routing region). Another example might be that the SIP container is used as an outbound Proxy. We might only want to do outbound services to our own users: If a user from another domain is bouncing over our server, then we could choose to send it to an application that would reject it, or we could proxy it on according to the next hop SIP rules, if we want to be nice. One scenario that the Application Router framework can help us with would be a case in which we have been running some application on our server and then want to break out to another server but also to get it back later and continue to execute some services before we are done. Imagine that we have run a call-screening application and then we want to forward the incoming INVITE SIP request to National Security Agency (NSA) SIP server. Maybe they want to monitor all SIP traffic so that they can wiretap. They would not trust us in running the service, so it would be hosted in their server park. With the AR framework, we can specify in the "SipApplicationRouterInfo" an external route pointing to the NSA SIP server. Then, for indicating that we want to resume the control, we can then use the "SipRouteModifier.ROUTE_BACK." The SIP container would then push two SIP "Route" headers on to the outgoing request.

```
Route: <sip:wiretap@nsa.gov>;lr
Route: <sip:oursailfin@sipservlet.net>;lr
```

The different scenarios can be many, so it is up to the application deployer to choose the tools and algorithms. The application router should be able to be totally stateless, and for that purpose we have the "stateInfo" serialized object that we can ask the container to store for us. Other than that, a simple properties file can be included in the AR ".jar" file so that the AR can retrieve the initial configuration. There is nothing preventing the AR from having its own database Back End (BE) or some other repository. It is a simple component that lives inside the application server, having its own class loader. It could have a GUI to be run time configurable or use JMX to communicate to some other entity. One thing to respect is not to modify the "SipServletRequest" object, because this could have serious consequences. Any modifications needed to the SIP request should be delegated to a helper application working on behalf of the deployed AR. This is a quite powerful pattern in which guards can be written. For example, if the AR wants to allow only authenticated users to execute its service chain, then when an unauthorized request is spotted, then the next application should be the "AuthApp," which is a mandated pattern in this use case. In this way only authenticated calls to service can be enforced by the AR and the helper "AuthApp."

8.7 Cluster Deployment

So far we have been dealing with single instance development, but what happens if your service gets popular and you want to add more servers? Configuring Sail-Fin in cluster mode is a bit more complicated, but the actual tasks you need to perform are quite straightforward. First of all, there is some architectural background information that should be mentioned in order for the practical steps to make sense.

We have already touched upon the central controlling entity in a SailFin deployment, the Domain Administration Server (DAS). So far, running on the single instance, the DAS has been used as "the" server. Now in a cluster setup, it is still possible to deploy applications, but there is a strong encouragement to avoid it by the responsible architect designers. The reason for this is that you really want to avoid overloading your administrative server. Deploying other administrative applications is good, but do not deploy the ones that take heavy SIP or HTTP traffic.

For every physical machine (e.g., a PC or a server blade), you need to configure at least one Node Agent. The Node Agent is a stand-alone process whose main task is to listen for commands from the DAS and keep the configuration synchronized between the DAS and its server instances. At same time, the Node Agent will keep track of server instances. If one instance should fail, it will then try to restart it. The Node Agent also provides a log file, where the success or any failures of its monitored server instances are reported. Then, in each machine handled by the Node Agent, there can be multiple server instances. As noted before, they need to have unique IP ports or run on different interfaces so that they do not clash in the OS. The server instances do not necessarily need to belong to the same cluster domain. Then the node agent needs to keep track of what "domain.xml" configuration file updates should be relayed to which server instance. For example, a system could look like DAS, Node Agent, and "cluster-a" and "cluster-b." Now, server instance "i1" is part of "cluster-a," while "i2" is part of the "cluster-b" domain. Even if it can appear funny to run like this on one machine, it makes perfect sense, because every SailFin domain is symmetric. What that means is that all applications need to be deployed on all server instances in a domain cluster. So, if you want two different applications but still want the high-availability functionality, this is how you need to configure your system.

So let's look at how a simple cluster running on one machine would look. For simplicity, we will define only one cluster, with two server instances. There is the single machine cluster deployment, as shown in Figure 8.2.

Figure 8.3 shows the same cluster, now on three machines.

In order to set this up, we need to do the following:

- Create a clustered configuration of a domain. (This would create the DAS.)
- Start the domain.
- Create a Node Agent.
- Start the Node Agent.
- Create the cluster (cluster-a).
- Create the server instance i1 in cluster-a.
- Create the server instance i2 in cluster-a.
- Start the cluster (cluster-a).

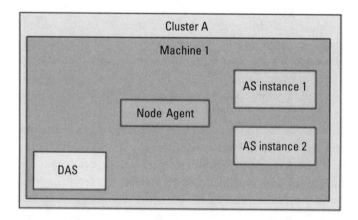

Figure 8.2 Cluster of one physical machine containing two server instances.

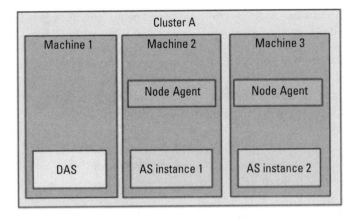

Figure 8.3 Cluster of three physical machines containing one server instance each.

- Change the log level for SIP.
- Deploy a SIP Servlet archive application.

Preferably, this is scripted, and we will provide a simple script doing this. The first step is to create a domain that runs in the clustered mode.

```
lib/ant/bin/ant -f setup-cluster.xml
```

The setup Ant script does the same thing as the "setup.xml" file but with the distinction that a clustered domain is created with all the possibilities that HA brings. The DAS is created and given the same ports as in the single node scenario.

```
bin/asadmin start-domain —terse=false —-passwordfile passfile domain1
```

This time the "start-domain" command takes additional parameters. That the statement terse equals "false" means that we do want a more extensive logging for the operation for a human reader. Setting it to "true" would produce minimal logging of the progress. Another parameter is given with the password file as input. This file needs to have three different passwords specified:

AS_ADMIN_PASSWORD=verysecret
AS_ADMIN_ADMINPASSWORD=verysecret
AS_ADMIN_MASTERPASSWORD=changeit

These are the ones we are going to use for our setup.

The next step is to create the node agent and start it.

```
bin/asadmin create-node-agent --terse=false --host 10.0.0.5
--passwordfile passfile a-n1
```

```
bin/asadmin start-node-agent --terse=false --passwordfile passfile a-n1
```

The new variables here are the host, which simply is the IP address of our machine. Then, at the end of the command, we specify the Node Agent name to be "a-n1" (Agent—Node 1: This way, if we should have multiple machines, the second would be "a-n2," but the naming convention has to be meaningful only for the system administrator). Beware of picking too-long names for the Node Agent's and server instances, since they are later concatenated into file directory structures. Shorter names make it easier to navigate later on.

Now for the creation of the cluster, which we name here "cluster-a." (So, following this notion, the next cluster we create would be "cluster-b.")

```
bin/asadmin create-cluster --terse=false --host 10.0.0.5 --password-
file passfile cluster-a
```

Next step is to create the two instances "i1" and "i2."

```
bin/asadmin create-instance --terse=false --host 10.0.0.5 --password-
file passfile --nodeagent a-n1 --cluster cluster-a server-a-n1-i1
```

```
bin/asadmin create-instance --terse=false --host 10.0.0.5 --password-
file passfile --nodeagent a-n1 --cluster cluster-a server-a-n1-i2
```

The last parameter specifies the instance name. For simplifying and making the orientation better, we append server "a," which is our cluster postfix, with "n1," which indicates what node agent is responsible, and finally "i1" or "i2" for identifying the instance. The reasoning behind this is the same as before: to simplify for the administrator of the system.

The next step is to start the cluster:

```
bin/asadmin start-cluster --terse=false --host 10.0.0.5 --password-
file passfile cluster-a
```

Now the cluster should be up and running. In some cases, there could be some warning-level logging while performing these operations, so to verify that everything is up and running, we could simply type a "netstat" system command. We never specified the ports to be used, and the "asadmin" command would default to the very first SIP post to be 35060 for UDP and TCP SIP connectivity. It simply prefixes the standard SIP 5060 port with the number 3. The TLS SIP server socket would be found on 35061. For some strange reason, the next instance "i1" has the ports in reverse order, so the TLS port is on 35062 and the UDP and

TCP on 35063. The logic behind it is hard to understand, but it is also possible to control the definition of the ports by properties in the "create-instance" command. However, by doing the "netstat," it is quite simple to figure this out, since according to RFC 3261, a SIP server always needs to listen on the same port for both UDP and TCP, while the TLS cannot share a port with any UDP service.

```
~ stoffe$ netstat -an|grep 3506
tcp46  0  0 *.35060      *.*          LISTEN
tcp46  0  0 *.35061      *.*          LISTEN
tcp46  0  0 *.35063      *.*          LISTEN
tcp46  0  0 *.35062      *.*          LISTEN
udp46  0  0 *.35060      *.*
udp46  0  0 *.35063      *.*
```

Here, in print, it is easy to spot the UDP ports: "i1" runs on 35060 and "i2" on 35063. Now let's try this out, but first we need to adjust the SIP logging.

In a clusterd domain configuration, the actual cluster name needs to be provided, so the setting of FINE logging looks something like this:

```
asadmin set "cluster-a-config.log-service.module-log-levels.prop-
erty.sip"=FINE
```

Now we need to look at the log for both "i1" and "i2."

The best way of running this on a UNIX system is to tail the two files.

If you have followed the previous instruction then the logs would be located at

```
sailfin/nodeagents/a-n1/server-a-n1-i1/logs/server.log
```

and

```
sailfin/nodeagents/a-n1/server-a-n1-i2/logs/server.log
```

As an alternative, the logging level can be changed with the Web admin GUI as shown in Figure 8.4.

Note that the GUI menus look different for a clustered domain!

Now let's deploy the "SimplestApplication" again:

```
bin/asadmin deploy --target cluster-a SimplestApplication.sar
```

Now we need to generate some SIP traffic. Once more, we can utilize the SIP Test Agent in NetBeans. This time we need to change the port that we are targeting traffic at to 35060. Create a SIP message and send it. The SIP 200 OK should be received, and the log of "server-a-n1-i1" should show the trace that the SIP MESSAGE went in and the 200 OK was sent back.

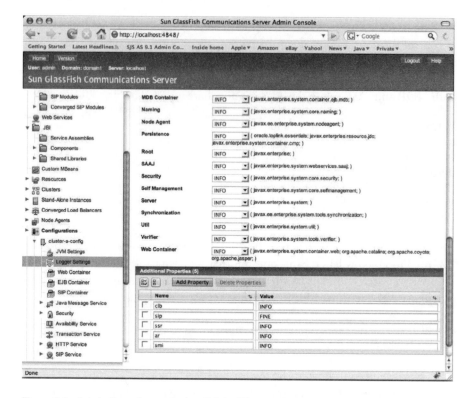

Figure 8.4　Admin Console now looks slightly different when in cluster mode.

The next step is to go back to the SIP Test Agent and change the destination port to 35063. Now we will send again, and this time the SIP MESSAGE should apear in "server-a-n1-i2 log."

Note　So far this has been pretty straightforward administration that is in line with what GlassFish provides. One good blog entry on this topic was made by Kedar, who is the admin lead at Sun [2]. This also links to the official GlassFish version 2 (GFv2) admin pages, where you can find more information on how to configure SailFin. The only Sail-Fin specific part is that the SIP container is brought up together with the JEE 5 Glass-Fish server.

Also, it is possible for the Web admin GUI to transform a single instance domain created by the "setup.xml" file and into a cluster domain setup. Since the steps in creating a new cluster domain are quite simple and give much better control, we encourage you to follow the described procedure.

8.7.1 Load Balancing and IP Sprayers

As you saw in the previous example on how to deploy a cluster domain, the result is that you have multiple instances running on the same or different IP addresses and port combinations.

To have a common understanding, let us set the definitions on what is a load balancer and what is an IP Sprayer. The IP Sprayer,works on the IP address level, as the name implies. Once an IP address or a domain name is provided, it would then keep track of the server instance in the server park and forward the traffic. A common algorithm is to use round robin so that every new request is sent to one of the hosts. The next request is sent to the next on the list. So, in our previous example, we would have one address that would be first-time forwarded to port 35060, and the second time around, to 35063. Then the algorithm would start over with port 35060. This is probably a normal router or a Network Address Translater (NAT) device, or we could use a module like "netfilter/iptables" on UNIX.

As for the load balancer, this is a more advanced extension of the IP Sprayer and also inspects the protocol layer. In our case, it should be capable of understanding SIP and HTTP protocols. These products are often much more expensive, but at the end of the day they are quite easy to set up. The drawback is that they are not part of the SailFin cluster, so they need to discover the availability and the current cluster shape of the AS running behind it. The two systems, the load balancer and the SailFin cluster, could have different opinions on who really is up and running. So, a misbehaving cluster could cause some pain. The other drawback it has is that all the SIP and HTTP protocol extensions now need to be implemented in both places. If the load balancer does not support Comet for HTTP or the SIP REFER method, then its inspection and balancing will be taken on false grounds. Examples of popular load balancers include Big IP F5 and Cisco PIX among others. The functionality of inspecting protocols is sometimes referred to as an Application Level Gateway (ALG).

So for a full-scale deployment, we would like to provide only one address where the provided service can be found. Basically, there are three options we can exploit:

- ALG Load Balancer + SailFin no Converged Load Balancer;
- IP Sprayer + SailFin Converged Load Balancer;
- DNS (NAPTR+SRV) + SailFin Converged Load Balancer.

The ALG deployment scenario when no CLB support is enabled as shown in Figure 8.5.

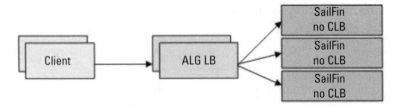

Figure 8.5   SailFin cluster behind ALG load balancer.

The CLB deployment where the IP Sprayer is randomly sending request to the various SailFin instances as shown in Figure 8.6.

The DNS load balancer deployment where each SailFin CLB FE is registered in the DNS, as shown in Figure 8.7.

The ALG deployment is quite traditional and straightforward. The ALG will make sure to which all SIP dialog-related information ends up on the same SailFin instance which the initial signaling was sent to. In more complex cases, such as a presence application in which all the subscriptions for a user can be spread out all over the cluster, there can be a significant performance hit. Espe-

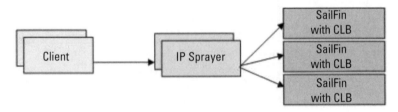

Figure 8.6   SailFin cluster behind simple IP Sprayer, load balancer in SailFin Cluster.

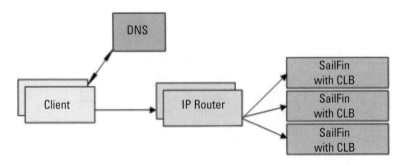

Figure 8.7   SailFin cluster behind DNS IP Sprayer, load balancer in SailFin Cluster.

cially when the cluster grows, the performance will drastically decrease. In some cases, the vendor provides hooks so that a more advanced load-balancing algorithm can be written and deployed. As mentioned earlier, this can be a quick way to deploy a service and as long as the traffic scenario is supported by the ALG—then it is all fine.

The next possible deployment is to use a much simpler and less expensive IP Sprayer. Then the ALG and load balancing will be taken care of by the CLB component in SailFin. There we have two possibilities. The default algorithm is called *user centric* and it will take the user part of the SIP Request URI to load the balance on. If that one is missing, then it will use the user part of the SIP URI in the SIP "To" header. There is also a notion for the "Originating" (as defined in JSR 289 [3]) call region, used especially for IMS deployments where this is specified in a lot more detail. In the originating case, it would first look at the user part of the SIP URI for the "P-Asserted-Identity" header, and if that one were missing, then the user part of the SIP URI in the SIP "From" header would be used. Since it is falling back to the SIP "From" header, this could also be used for a standard SIP deployment. The only criterion is that the SIP "Route" header that got the message to SailFin contain a URI parameter equal to "call=orig." In this case, the SailFin cluster has the knowledge of which servers are currently part of the cluster and can redirect traffic without delay.

If this is not the load balancing that is required for a particular service, then the first option is to change the actual string that is used for the load balancing. Instead of taking the user part, some other part of the message can be inspected and another string used. The other alternative is to replace the entire user (data)-centric algorithm with a custom-made one. It could look at the day-in-week or the actual load on the servers to accommodate the result. This is not much different than having to write ALG load-balancing rules. In this case, the development environment is in Java and is a matter of preference. Of course, all the other pros and cons mentioned previously still apply to separate the different solutions.

To change the string that the CLB uses for inspection, an XML file according to the Document Type Definition (DTD) can be located under the following:

```
SailFin/lib/dtds/sun-data-centric-rule_1_0.dtd
```

A sample of the file can be found in the "SJS Communications Application Server HA Guide" at [4] (look under the CLB chapter, "The Data Centric Rules File").

To deploy a new rules file, either the "asadmin" command or the Web admin GUI can be used:

```
bin/asadmin set domain1.converged-lb-configs.myclbcfg.converged-lb-
policy.dcr-file=dcr.xml
```

As for a total remake of the algorithm, there is no Service Provider Interface (SPI), but the CLB code is in the SailFin CVS, so careful modification can be easy or quite hard depending on what the new algorithm looks like. The code can be used as inspiration, and since both the CLB front end and back end are layers, they can be replaced as described above.

The last option available is to use DNS as the load balancer. The assumption is that the SIP clients consuming the service are compliant with RFC 3263 [4]. The SailFin SIP container is also compliant with RFC 3263 when it acts as a UAC, but in this scenario for load balancing, it would be acting as a UAS, B2BUA, or Proxy. Just to explain what this RFC is all about and what kinds of DNS queries would be sent, we will describe a full location scenario. There are multiple levels of load balancing, since the first choice when a SIP UAC wants to send a SIP message is to pick the transport. For this reason, the Name Authority Pointer (NAPTR) DNS record format RFC-2915 [6]) was defined. From the Request URI of the SIP message that is being sent the domain is extracted.

```
MESSAGE sip:stoffe@sipservlet.net SIP/2.0
```

Now the imaginary service we try to access is located in the "sipservlet.net" domain. To find out what SIP protocols are supported, the client needs to post a DNS NAPTR query.

```
OPCODE=SQUERY
QNAME=sipservlet.net,QCLASS=IN,QTYPE=NAPTR
```

The response from the server for a SailFin cluster could look like this:

```
IN 0 NAPTR 10 10 "s" "SIP+D2T"  "" _sip._tcp.sipservlet.net
IN 0 NAPTR 20 10 "s" "SIPS+D2T" "" _sips._tcp.sipservlet.net
IN 0 NAPTR 30 10 "s" "SIP+D2U"  "" _sip._udp.sipservlet.net
```

The fields for each row are defined in according classes from the RFC-2915 listed here:

```
Domain      Class      Preference      Service      Replacement
TTL         Type       Flags           Regexp
```

The client received three entries that state the preferences for the "sipservlet.net" domain. It is easy to spot that the domain supports SIP over UDP, SIP over TCP, and SIPS over TLS.

The most interesting part is the "Order" field. According to the RFC, the one with the lowest value should be chosen. In this example, the first line has a value of 10 in the order field, while the other two lines have 20 and 30.

The "Preference" field is not significant, because it comes into play only when there are two rows with the same number in the "Order" field. By configuring our DNS like this, we state that, if the client does not know what protocol to use, then we mandate it to use TCP. This is the first step where load balancing is done. There is additional information contained in the NAPTR record. The flag states "s," which indicates that the "Replacement" field contains a pointer to a Service (SRV) DNS record.

SRV is defined in RFC-2782 [7] and specifies the format of the service records. Now the client needs to take the result from the NAPTR lookup and do an SRV DNS lookup. As an alternative, if the client knows that it must use SIPS over TLS, it can then skip the NAPTR lookup and do the SRV straightaway.

```
OPCODE=SQUERY
QNAME=_sip._tcp.sipservlet.net,QCLASS=IN,QTYPE=SRV
```

The SRV query looks quite similar to the NAPTR one, and it is quite straightforward. Now let's look at the answer.

```
_sip._tcp.sipservlet.net 0 IN SRV 1 1 35060 mysailfin.sipservlet.net
                         0 IN SRV 1 1 35063 mysailfin.sipservlet.net

mysailfin.sipservlet.net 0 IN A 10.0.0.5
```

The fields for each row are defined in according classes from the RFC 2782, listed here:

```
_Service._Proto.Name TTL Class SRV Priority Weight Port Target
```

So now to the answer: there are two SRV record rows received. Note that they are reflecting the SailFin cluster that was built in the previous section. Both have the same "Priority," so the client should first use one, and then, for the next time, it will have done a DNS query and should use the alternate one. Both rows also have the same "Weight" factor, and that indicates only that both should be loaded. If the first one has a weight of 3, then the client should send three times as much SIP traffic there as to the second entry. Normally, this is done when one piece of hardware is much more powerful than the other, but in reality, most server parks have equal hardware, making this feature redundant. In our case, it is the same machine, so the values of 1 and 1 make perfect sense. The SRV "Target" is pointing to "mysailfin.sipservlet.net," and the client needs to get an IP Address. To minimize the amount of DNS queries, the provider has chosen to put the DNS record for the "mysailfin.sipservlet.net" into the SRV answer. This is perfectly legal, but if it were not there, then the client would be forced to do one additional lookup to finally receive the IP to use. It already knows the port from the SRV query. In this example, the client gets back the IP address of 10.0.0.5.

As you can see, the client now has a lot of information. The SIP request should be sent over TCP to IP 10.0.0.5 and port 35060. A side effect is that, if the server should happen to go down, it also knows that it could try TCP to IP 10.0.0.5 and port 35063, which runs the same service that it wants to access.

The result is that DNS NAPTR and SRV can both give load balancing and high availability. It does not cost any investment in load balancing hardware, but it requires that your DNS provider allows for provisioning of its DNS with the required records. In a larger setup, the system would most probably contain a specific DNS subdomain and could host that zone in a service-specific DNS. A more likely scenario is that only one SRV record would be returned, since all the Sail-Fin instances would run on port 5060, and then the "mysailfin.sipservlet.net." A record answer would contain multiple entries. One row per specific SailFin cluster member IP address would be returned. This list can be rotated so that the topmost one is different for every new query until the list wraps around.

The catch with this solution is that all the IP addresses returned have to be routable. In this example, we used IPv4 addresses, and that can be a limitation. Don't forget there might be need of a firewall and some denial-of-service detection and exclusion of misbehaving clients. One possible solution for the IPv4 limitation would be to use IPv6, since it doesn't have the address range as a limiting factor. As for the firewall, it all depends on the service that is being provided, so it is hard to make recommendations.

8.7.2 ENUM

Even if the DNS NAPTR and SRV is not the chosen location mechanism of your deployment, it is still extremely important to understand how it works. As mentioned previously, the SailFin SIP container will try to do this automatically if an Address of Record (AOR) is provided for the next SIP hop after SailFin. There is one more use case in which a DNS server would be queried automatically by SailFin. This case occurs when the next route contains a TEL URL. This is only interesting when building Plain Old Telephony System (POTS) replacement with a SIP Voice over IP (VoIP) solution.

Let's assume that I used to have a telephone number +1 555 1234 567. Now, the TEL URL in the SIP request header would look like this:

```
INVITE tel:+15551234567 SIP/2.0
```

Note It could also be indicated in the following way: sip:+15551234567@ sipservlet.net;user=phone.

This request going out from SailFin is quite hard to route, because we have little knowledge of the target destination. For this reason, the ENUM standard was specified (RFC 3761) [8]. The RFC also defines a mechanism for the client to find the ENUM service with NAPTR and explains how the phone number should be stripped to form the query address record that should be used in the DNS ENUM query. First, all nondigit characters are removed, and then the number is reversed with a "." sublimiter between every digit.

```
+15551234567 becomes 7.6.5.4.3.2.1.5.5.5.1.e164.arpa
```

The domain "e164.arpa" is included only for the purpose of facilitating telephony number translations. Since it is used as my number but then I chose to migrate to a VOIP account, there would be a corresponding SIP address pointing to me.

A result of the ENUM query could look like this:

```
NAPTR 10 100 "u" "E2U+sip" "!^.*$!sip:stoffe@sipservlet.net!"
```

All the friends that have my old POTS number would be able to still use it, while to any new friends I would hand out the SIP URI of "sip:stoffe@sipservlet.net" instead.

Note OK, maybe not Grandma, since she would not know what to do with a SIP URI.

Not only does this happen automatically for the next SIP hop in SailFin, but the SIP container also provides a proprietary interface for looking up a TEL URL and mapping it to a SIP URI. Under the hood, this uses the same DNS ENUM client, but the TEL URL could be delivered in any other SIP header, or it might be provisioned and looked up from a Structured Query Language (SQL) database.

The class "org.glassfish.comms.api.telurl.TelUrlResolver" is the proprietary extension, and it contains two methods:

```
public SipURI lookupSipURI(URI uri) throws IOException,
TelUrlResolverException;

boolean isTelephoneNumber(URI uri);
```

The "isTelephoneNumber" takes a "javax.servlet.sip.URI" object and checks whether it's a TEL URL that is syntactically correct or a SIP URI with a "user=phone parameter" that has a numeric user URI part. The "lookupSipURI" method has the same input, but if the URI were to apply to all the rules, it would then result in a ENUM DNS query, and the answer would be returned as javax.servlet.sip.SipURI interface instance. Note that the ENUM query could also return e-mail record lines or H323 (former signaling standard competing/

being replaced by SIP) records (as shown in the examples of RFC 3761 [8]). Sail-Fin would return the first valid SIP entry encountered by a call to this function.

So how to get a reference to this interface?

The simplest way is probably to do this in the "init()" of the SIP Servlet that wants to do ENUM lookups.

```
@Resource SipFactory sf;

@Override
public void init(ServletConfig config) throws ServletException {
    super.init(config);
    ServletContext ctx = config.getServletContext();
    TelUrlResolver telResolver = (TelUrlResolver)
ctx.getAttribute(TelUrlResolver.CONTEXT_ATTRIBUTE_NAME);
    TelURL tel = (TelURL) sf.createURI("tel:+1-555-1234 567");
    SipURI sip = telResolver.lookupSipURI(tel);
    log("Result of ENUM ="+sip.toString());
}
```

If everything was set up correctly, it should log "Result of ENUM = sip:stoffe @sipservlet.net." The provisioning of the ENUM records is outside of the scope of this book, but there are some publically available ENUM servers that can be used for testing this out. One important configuration aspect of the SailFin DNS client that probably needs to be configured is specifying what DNS servers to post queries to. It is a JVM–D flag, then it can be edited in the "domain.xml" configuration file with the admin Web GUI or by "asadmin" command when the server is stopped.

```
-Ddns.server10.0.0.5
```

With that JVM flag setting, the host 10.0.0.5 would be queried on the standard DNS UDP port 53.

All these procedures are in line with the administration procedures shown in Chapter 7.

References

[1]   Rosenberg, J., et al., "SIP: Session Initiation Protocol," RFC 3261, Internet Engineering Task Force, June 2002.

[2]   Kedar's blog on GlassFish administration, http://blogs.sun.com/bloggerkedar/entry/how_ das_communicates_with_node.

[3]   SIP Servlet Specification, Version 1.1, JSR 289, Java Community Process, August 2008.

[4]   SailFin Cluster Admin Guide, http://docs.sun.com/app/docs/coll/1343.8.

[5]   Rosenberg, J., and H. Schulzrinne, "Session Initiation Protocol (SIP): Locating SIP Servers," RFC 3263, Internet Engineering Task Force, June 2002.

[6]   Mealling, M., and R. Daniel, "The Naming Authority Pointer (NAPTR) DNS Resource Record," RFC 2915, Internet Engineering Task Force, September 2000.

[7]   Gulbrandsen, A., P. Vixie, and L. Esibov, "A DNS RR for Specifying the Location of Services (DNS SRV)," RFC 2782, Internet Engineering Task Force, February 2000.

[8]   Faltstrom, P., and M. Mealling, "The E.164 to Uniform Resource Identifiers (URI) Dynamic Delegation Discovery System (DDDS) Application (ENUM)," RFC 3761, Internet Engineering Task Force, April 2004.

9

SIP Servlet Client Programming

The majority of SIP Servlet development and deployment occurs in the network while the SIP Container is acting as an application server. Even then, the trend is toward micro containers, but the technology is not there yet. Today it is hard to find a SIP Servlet container that can run on a mobile phone or any other limited footprint device. The SailFin SIP Servlet container, which has been used throughout this book, is based on the Sun GlassFish Version 2 (V2) JEE container. The GlassFish V3 is being built with a microcontainer core, which will allow components to have a much smaller memory footprint. Such a concept is probably two to three years away. JBoss is also developing a similar microcontainer architecture, so it is likely that it will be possible in the future to program SIP Servlets even on a limited footprint device.

Today, it is possible to take a fatter JSR 289 SIP container and build applications for a PC with graphical interface fronting the end user, in which the SIP container is used as a SIP message stack. Since the footprint would end up close to 100 Mbytes, it does not provide a good production solution. There are other SIP stacks, such as the NIST Jain SIP, that are more adapted to be embedded for client development. Programming Jain SIP is a totally different paradigm, and it requires a little more SIP knowledge of the programmer.

What is a more likely deployment, and that which follows the current trend of web clients, is to use a browser-based client approach. A browser is the user Interface, and then on the server side, there is interaction with the Web server for a classic Model—View—Control (MVC) pattern. Since most of the JSR 289 containers in the industry are bundled with JEE, the natural choice is to use JavaScript and AJAX on the browser while using HTTP Servlets and higher level packages like JSP and JSF (Java Server Faces). Since the core of JSP/JSF is HTTP Servlets, the converged container capabilities of the JSR 289 can be exploited.

Another alternative for writing HTTP-based clients is to use the Representational State Transfer (REST) [1]. There is an ongoing JSR in JCP (JSR 311) where JAX-RS is being defined. SIP Servlets can be used to implement a SIP UA client where all of its state data, such as a buddy list, conference state, initiation of a call/message, and presence information, could be accessed using the REST patterns. The browser-based client can then use JavaScript to access the REST objects over HTTP and when the object is received in an "HttpResponse" that the JavaScript can then render within the browser.

For this purpose, the "XMLHttpRequest" object in JavaScript is used to send GET/POST/DELETE commands to change the state of the SIP UA running on the SIP container. The biggest disadvantage that HTTP and a browser-based approach have is that HTTP is request/response-based protocol while in SIP the User Agent is a client (UAC) and a server (UAS) at the same time.

The problem can be illustrated in a simple chat application: Say the client wants to send a message to the person he is chatting with; we just simply wrap the text that he wants to send in a "HttpRequest." On the converged server side, the body is simply picked up and a new SIP MESSAGE request is created. This all runs smoothly. The problem occurs when the person we are chatting with wants to respond. The message arrives at the same SIP Servlet, but now, since HTTP is a request/response-based protocol, there is no way for the SIP container to send this message in a "HttpRequest" targeted for the browser. For this purpose, the most common programming pattern is that the browser regularly polls for new messages intended for it. This consumes a lot of resources at the client, in the network, and especially at the server. How often the connection is polled is difficult to determine and depends on how many clients a converged SIP and HTTP server is able to handle at any one time.

There are nonstandard techniques called asynchronous AJAX or COMET that keep one HTTP connection open at all times. The HTTP Servlet container simply does not answer on the "HttpRequest" until it has something for the client. At the same time, if the client wants to send something, it simply opens a new HTTP connection and, from then on, behaves as a standard entity, sending an "HttpResponse" without blocking. One example of such an application that is using this pattern is Google Mail. When an e-mail is received, it is instantaneously received in the browser and rendered by JavaScript. This gives far better responsiveness but also requires a proprietary extension on the server side. Changing the application server vendor is also harder, because this part is not standard.

Another alternative that might be interesting for developing mobile clients is to use Java Micro Edition (Java ME) and JSR 180. When using JSR 180, the Java ME Mobile Information Device Profile (MIDP) connector architecture can be used to generate and receive a SIP message. This is quite low level, but it allows for any kind of message to be generated or consumed. At the same time, the Java ME GUI classes can be used to make the user interface.

Why would one choose to program SIP all the way to the terminal when there are browser-based solutions? At the time of this writing, it is about penetration into the marketplace. JSR 180 is out there prebundled on many mobile phones because it is part of the JSR 248 MSA (Mobile Service Architecture). At the same time, the more limited mobile browsers do not support asynchronous AJAX/COMET, so there would be a lot of polling, and depending on the mobile subscription, it could be very slow and costly for the end user of such a client.

There is no silver bullet, but one has to evaluate and pick the architecture that suits the intended deployment. This chapter will provide an overview explaining what this kind of programming is all about. It could very well be that you will never need to make a client because softphone clients, hardware SIP phones, or terminal adapters are used in your deployment. However, if you really want to do innovative client applications, there is a fair chance that you will have to do client programming too.

9.1   Writing HTTP Servlet-Based Client

Even if you are not going to do an HTTP Servlet-based client, do not skip this section. There is a fair chance that you will have to build one for administrative purposes. First of all, because of convergence, this is a very efficient way to build a client, since both the SIP parts and the HTTP parts execute in the same JEE application server.

To create a simple example, we could write a Web-based chat client. The client could have a simple HTML page, JSP page, or even a JSF page where we can enter a text message, a SIP target URI, and a send button. Inserting JSF tags updates the objects and makes them accessible on the server side, where we would use the SIP Servlet "SipFactory" helper interface to create a new SIP message. Since the ".sar" file format allows for HTTP components, we simply need to include a "web.xml" file and the required "faces-config.xml." At the same time, the JSF ".jar" files need to be bundled in the WEB-INF "lib" directory so that the dependencies are satisfied.

Let's start with the simplest case, in which we interact with an HTML page.

9.2   Using Asynchronous HTTP

To write a client using asynchronous HTTP requires a choice of server in which the technology is available. Since we have been using the SailFin open source SIP container, the example will use the Grizzly COMET support, which is available in the HTTP container. First, we will do the same example with the chat application, but there are also "add-on" frameworks to hide more of the details. One of the frameworks that works fine with SailFin and COMET is ICEfaces (www.icefaces.org). The ICEfaces team is attempting to keep the framework portable over multiple applications servers, so if you want more portability, ICEfaces is an option.

First, the COMET support needs to be enabled in the SailFin HTTP container as follows:

```
<SailfinDir>/domains/domain1/config/domain.xml
<http-listener acceptor-threads="1" address="0.0.0.0" blocking-
enabled="false" default-virtual-server="server" enabled="true"
family="inet" id="http-listener-1" port="8080" security-
enabled="false" server-name="" xpowered-by="true">
    <property name="accesslog"value="${com.sun.aas.instanceRoot}
/logs/access"/>
    <property name="cometSupport" value="true"/>
</http-listener>
```

The line in bold text "<property name="cometSupport" value="true"/>" is added to the default domain.xml file.

Basically, the container will allow an "HttpRequest" to be blocked and stored away in a "HashTable." In normal operations, when an "HttpServlet" does not answer on a request, then the container would automatically respond with a 500 error code. This is all that has to be done when it comes to configuration.

Let's create a HTML page for collecting the appropriate chat information. We want to have an input field specifying who we want to chat with, a larger text field containing the chat history, and a field where input can be collected for the latest message to submit. We also want a send button.

To get the traditional chat client look, the HTML code would look something like this:

```
<html>
  <head>
    <title></title>
    <meta http-equiv="Content-Type" content="text/html; charset=UTF-
8">
    <script>
      var hist = new Array();
      var to = "";
      var cometReq;

      doGet();
.
.
.
    </script>
  </head>
<body>
```

```
    <form name="sendForm">
      Target SipURI : <input type="text" name="target"
value="sip:localhost:5070" size="46" />
      <br/>
      History : <textarea name="history" readonly="true" cols="40"
rows="10">
      </textarea>
      <br/>
      Message : <input type="text" name="message" on
onkeypress="return submitOnReturn(event);" value="" size="52" />
      <input type="button" onclick="sendMessage();" value="Send"
name="sendButton" />
    </form>
  </body>
</html>
```

The HTML rendered in Firefox will result in a nice GUI as in Figure 9.1. If you look carefully, you'll notice there are some JavaScript methods executed in the HTML code above. The "onkeypress" event that calls the "submitOn-Return(event)" is simply a convenience method that monitors characters that are inserted into the message input text field when the return key is pressed and the message would be submitted. What happens behind the scenes is that the second JavaScript method "sendMessage()" is called. Here is how the method looks:

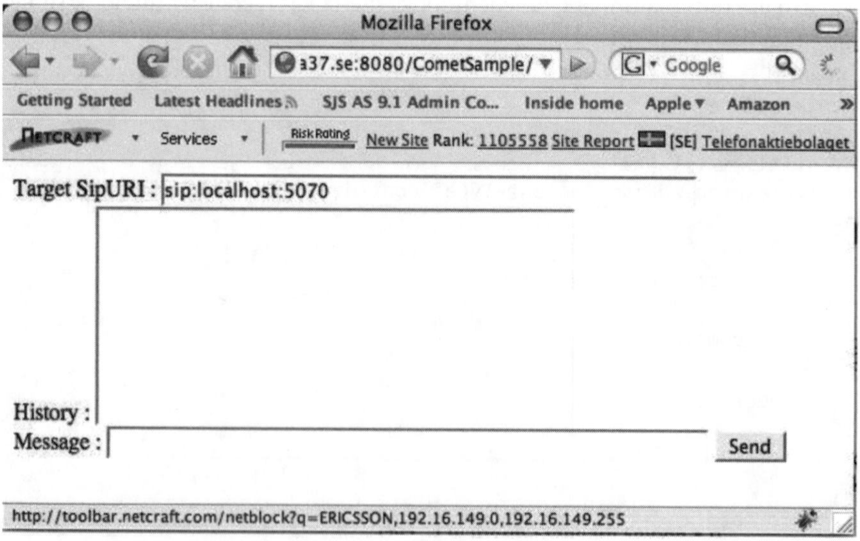

Figure 9.1   HTML Form based client interface.

```
function sendMessage() {
        var req = new createXMLHttpRequest();
        var target = document.forms[0].elements["target"].value;
        var text = document.forms[0].elements["message"].value;
        var msg = "target="+target+"&message="+text;

        req.open("POST", "CometServlet", false);
        req.setRequestHeader("Content-type", "application/x-www-
form-urlencoded");
        req.setRequestHeader("Content-length", msg.length);
        req.send(msg);

        addToHistory("me: "+text);
        var start = target.indexOf(':');
        var end = target.indexOf('@');
        to = target.substring(start+1,end);
        document.forms[0].elements["message"].value = ""; //Clear it
};
```

First of all, we create a "XMLHttpRequest," and that is done in a helper method because older Microsoft browsers instantiate the object differently. The majority of the newer browsers simply let you call a new "XMLHttpRequest()." The next important part is to call open on the request; in this example, we are using the "POST" method. Then the second argument is the Servlet path, and the last parameter is whether the call should be asynchronous. The reason for stating the Servlet path to "CometServlet" is simply that the HTML page will be packaged in the same archive as the Servlets, so there is no need to provide full paths. Since posting a message is a synchronous activity, the third parameter to open is "false." Some HTTP headers are set to inform the server that it is the form data format (HTML Form) and also the length of the entire string. Finally, the data is sent using the "send(msg)" command. Note that the message contains the target and the text to be submitted. The remainder of the "sendMessage" function stores the sent message in the history and clears the input box so that a new message can be entered.

We have covered the sending part of the example. You might have noted that in the beginning of the HTML declaration, when the script diction was declared, there is a call to "doGet()." This is another JavaScript function that enables the COMET long polling HTTP connection to become established. The "doGet" method is called the first time the browser reads the page.

```
function doGet() {
        cometReq = new createXMLHttpRequest();
        cometReq.onreadystatechange = changed;
        cometReq.open("GET", "CometServlet", true);
        cometReq.send(null);
};
```

```
function changed() {
        if (cometReq.readyState==4)
        {// 4 = "loaded"
          if (cometReq.status==200)
          {// 200 = OK
            receiveMessage(cometReq.responseText);
            doGet();
          }
          else
          {
            alert("Problem retrieving XML data");
          }
        }

}
```

The "doGet" also creates a "XMLHttpRequest," but this time it is an HTTP "GET" that is targeted toward the same Servlet path. The big difference is that this time the third parameter to the open method is "true," enabling asynchronous HTTP. The other interesting part is that a callback for the "onreadystatechange" is set to call the changed function. The "changed" function is declared, and it waits to receive a 200 response that would call "receiveMessage()" method. Part of the processing of receiving a message is to extract the text from the HTTP response body and simply add as the latest entry to the history window, and then the "doGet()" function to keep an active HTTP connection. Remember that the GET is parked at the server side, so we need at all time to have an ongoing GET request, since we never know when the other chatting part will send something to us.

Let's now look at the server side to see how this POST and GET are received. To make it simpler, there is a JavaBean object holding information about our client and how SIP should be sent.

```
public class SipUser {

  protected String SipIdentity;
  protected String user;
  protected String password;
  protected String outboundproxy;
  protected String realm;
```

These are the associated bean fields with their corresponding "getter" and "setter" methods. In the example archive, these are bootstrapped from the "web.xml" file init parameters. If one would like to make better use of this example, probably there could be an initial configuration and authentication page on which a user would log in, instead of statically setting the parameters using the "web.xml" file. Here we will just omit that to focus on the important parts of the COMET communication. Here is the class declaration and the init function:

```
public class CometServlet extends HttpServlet {

  @Resource SipFactory sf;
  static String contextPath = "CometServlet";
  ServletContext ctx = null;

  @Override
  public void init(ServletConfig config) throws ServletException {
    super.init(config);
    CometEngine e = CometEngine.getEngine();
    CometContext c = e.register(contextPath);
    c.setExpirationDelay(3600*1000);
    ctx = config.getServletContext();
    //For now lets store the user sip id in the ctx
    //This is picked up by StartupListener
    String sipId = config.getInitParameter("sipIdentity");
    String authUser = config.getInitParameter("authUser");
    String password = config.getInitParameter("password");
    String realm = config.getInitParameter("realm");
    String outboundProxy = config.getInitParameter("outboundProxy");

    ctx.setAttribute("sipuser",
            new SipUser(sipId,authUser,password,realm,outbound
Proxy) );
  }
```

In the Grizzy framework, there is a singleton class where the bootstrapping is done.

```
CometEngine engine = CometEngine.getEngine();
```

The engine provides the entry point for the COMET support. In order to tell the HTTP container that this application is a COMET application, the URL that the client will identify itself with has to be registered.

```
CometContext cometContext = engine.register("CometServlet");
```

Now, when the "CometContext" receives a timeout, it can be set, since we do not want to wait forever. If there is nothing to send from the server to the client, it is a good programming practice to release the request. The server can then send a 200 response to the blocked "HttpRequest," and the client can determine whether another attempt to maintain a COMET connection should be performed.

```
cometContext.setExpirationDelay(3600 * 1000);
```

This statement would hold the COMET connection up for an hour before timing out.

The last thing in the init is to create the "SipUser" JavaBean, populate it, and store it in the Servlet context so that both HTTP Servlets and SIP Servlets can access it.

Now let's look at how the POST with the message that we want to send is handled:

```
@Override
  protected void doPost(HttpServletRequest request, HttpServlet
Response response)
  throws ServletException, IOException {
    SipUser user = (SipUser) ctx.getAttribute("sipuser");

    String target = request.getParameterValues("target")[0];
    String msg = request.getParameterValues("message")[0];

    SipApplicationSession sas = sf.createApplicationSession();
    SipServletRequest req = sf.createRequest(sas, "MESSAGE",
user.getSipIdentity(), target);
    req.setContent(msg,"text/plain;charset=UTF-8");
    req.pushRoute(sf.createAddress(user.getOutboundproxy()));
    req.getSession().setHandler("authServlet"); //In case of 407
    req.send();
  }
```

First, we decode the target that we want to send the message to and also the actual text message to send. Then, from within the HTTP Servlet, we use the injected "SipFactory" helper interface instance to first create a "SipApplication Session" interface instance and then to create a new SIP MESSAGE request. We pick our identity from the JavaBean configured in "init()," and for the SIP "To" header part of the new message, we use the target decoded from the posted data. Now we set the message received and set the content type to "text/plain." Pushing the outbound proxy URI is due to the fact that often SIP service providers use that deployment model, but also it is easier to use a fake domain that you might not be in control of but want to "borrow." Let's say that you have two users, "sip:alice@ericsson.com" and "sip:bob@ericsson.com," but in reality you want to send them to "sip:127.0.0.1:5060." The "setHandler" statement is just putting a UAC authentication Servlet (see Chapter 2, Section 2.3, "Security") that was described earlier. If a SIP proxy server requires authentication, then the Servlet would create the digest challenge in an appropriate manner based on the values from the "SipUser" bean. Finally, we send the message using the "send" method.

Now that the sending is completed, let's look at the receiving end that handles the incoming messages. For that purpose, another SIP Servlet is created that, on init, would register itself and also receive any incoming SIP MESSAGES based on that registration.

```
@Override
public void init(javax.servlet.ServletConfig config) throws
                                  javax.servlet.ServletException {
    super.init(config);
    ctx = config.getServletContext();
    try {
      List<SipURI> oi = (List<SipURI>)
                ctx.getAttribute("javax.servlet.sip.outbound
Interfaces");
      for (SipURI uri : oi) {
          if( "udp".equalsIgnoreCase( uri.getTransportParam() ) ) {
            myContact = uri;
          }
      }
      sendRegister();

    } catch (Exception ex) {
      log("Error", ex);
    }
  }

  protected void sendRegister() throws IOException, ServletException {
    SipUser user = (SipUser) ctx.getAttribute("sipuser");

    //Register the user
    SipApplicationSession sas = sf.createApplicationSession();
    SipServletRequest req = sf.createRequest(sas, "REGISTER",
                user.getSipIdentity(), user.getSipIdentity());
    req.pushRoute(sf.createAddress(user.getOutboundproxy()));
    req.setHeader("Contact", "<"+myContact.toString()+">" );
    req.setExpires(3600);
    req.getSession().setHandler("authServlet");
    req.send();
}
```

In the init method, this Servlet simply picks one of the SIP container interfaces that uses UDP and stores it away. Then, in the register method, a registration is made, also setting the auth Servlet in case a 401 is returned.

When a registration is completed, the other part that we chat with can send a message.

```
@Override
public void doMessage(SipServletRequest req) throws IOException {
    SipServletResponse resp = req.createResponse(200);
    resp.send();

    CometEngine e = CometEngine.getEngine();
    CometContext c = e.getCometContext(CometServlet.contextPath);
```

```
if( req.getContentType().equals("text/plain"))
  c.notify(req.getContent().toString().trim());
}
```

First of all, a 200 response is sent to signify that we have received the message. Then the "CometContext" is retrieved. In this example, we have only one COMET context, but if we were to build a fully fledged client, we would have one context per browser and user, making it a multiuser service. Then, if we indeed receive a plain-text message, we would notify the COMET context of the message arrival.

To complete the sample, we need to look at what has happened to the HTTP GET message in the "HttpCometServlet" and how the notify() method call is propagated.

```
@Override
protected void doGet(HttpServletRequest request, HttpServletResponse
response)
  throws ServletException, IOException {
    CometEngine e = CometEngine.getEngine();
    CometContext c = e.getCometContext(contextPath);
    ChatHandler h = new ChatHandler();
    h.attach(response);
    c.addCometHandler(h);
}
```

The next task is to register a "CometHandler" object to be triggered that will handle events for this "CometContext."

```
context.addCometHandler(handler);
```

The Handler interface defines the following methods:

```
public interface CometHandler<E> {

  /**
   * Attach an instance of E to this class.
   */
  public void attach(E attachment);

  /**
   * Receive <code>CometEvent</code> notification.
   */
  public void onEvent(CometEvent event) throws IOException;

  /**
   * Receive <code>CometEvent</code> notification when the underlying
   * tcp communication is started by the client
   */
  public void onInitialize(CometEvent event) throws IOException;
```

```
/**
 * Receive <code>CometEvent</code> notification when the underlying
 * tcp communication is closed by the <code>CometHandler</code>
 */
public void onTerminate(CometEvent event) throws IOException;

/*
 * Receive <code>CometEvent</code> notification when the underlying
 * tcp communication is resumed by the Grizzly ARP.
 */
public void onInterrupt(CometEvent event) throws IOException;

}
```

The most important method is the "onEvent()." There are six types of events that the "onEvent()" will be triggered on: INTERRUPT, NOTIFY, INITIALIZE, TERMINATE, READ, and WRITE. The one we are interested in is the NOTIFY event, which gets triggered when the client performs an HTTP POST to a "CometContext." It is simply intercepted, and a new SIP request is created with the help of the SIP Servlet "SipFactory" Helper interface.

```
public void onEvent(CometEvent ce) throws IOException {
      if( ce.getType() == CometEvent.NOTIFY ) {
        PrintWriter pw = resp.getWriter();
        pw.write(ce.attachment().toString());
        pw.flush();
        ce.getCometContext().resumeCometHandler(this);
      }
}
```

Finally, the last line calls "resume" "CometHandler." It basically allows the client to send an additional message when the user chooses to GET a new chat message.

Now the code is completed, and since the SIP Servlets are annotated, there is no need for any updates of the "sip.xml" configuration file. Just for reference, the "web.xml" file contains the bootstraping of the "SipUser" bean and also the declaration of the HTTP "CometServlet," together with the default "index.html" page that contains the form and JavaScripts. The "CometSample.sar" file can now be deployed, and for another communicating part, the X-Lite client has been used together with a free Internet SIP account.

A nice task for exploiting this technology further, but which is outside of the scope of this book, would be to introduce presence to this sample, making it even better.

Note A majority of the COMET SailFin supports are implemented in the Java package "com.sun.enterprise.web.connector.grizzly.comet."

9.3 Using ICEfaces

As mentioned previously, the ICEfaces framework (www.icefaces.org) could be used as an alternative to make an interaction similar to the one using COMET just described. The ICEfaces framework can be included as transparent glue between the JSF pages and the asynchronous HTTP COMET support. ICEfaces supports a majority of JEE servers and frameworks, and it provides a higher level of abstraction than asynchronous HTTP. It also provides tag libraries and JavaScript libraries so that the client side development of rich Internet applications is easier and faster.

In the following sample, we will exploit only the JSF COMET integration as a comparison to the previous COMET example. The simplicity is in the fact that a standard JSF bean can be developed without any knowledge of the asynchronous Web model. The only glue code that needs to be added is on the JSF bean when loading the page, an instance of the ICEfaces session renderer gets instantiated. When the JSF bean is updated, a call to render the session needs to be issued using the ICEfaces framework. Everything else in the code is standard JSF framework programming.

Note   The actual "SessionRenderer" interface in ICEfaces is still considered experimental and can change in the near future, but the concept will remain.

When a "ConvergedSipProject" is created in NetBeans, JSF framework also should be added. It then generates a "faces-config.xml" and includes the necessary libraries. Some ".jar"s from ICEfaces (1.7.2 was used) need to be added and put under WEB-INF library:

- Icefaces.jar;
- Icefaces-facelets.jar;
- Commons-fileupload.jar;
- Commons-logging.jar;
- Backport-util-concurrent.jar.

The SIP code is accepting SIP MESSAGES (also registrations), and it is enough just to send a message to join it. The user field in the SIP Request URI is used to choose what chat forum a message is sent on. The SIP "From" header will be stored, and for any posted message, all participants that have sent a message will receive the incoming messages.

Instead of doing another client, we will now do a Chat History monitor that is updated in real time (with the help of COMET). So, as soon as the chat server receives the message, it will be updated on browsers monitoring the chat.

For this purpose a JSF bean and a JSP page are created.

```java
public class HistoryBean {
 private Vector<HistoryEntry> history = new Vector<HistoryEntry> ();
 private int maxSize = 15;

 public int getMaxSize() {
  return maxSize;
 }

 public void setMaxSize(int maxSize) {
  this.maxSize = maxSize;
 }

 public HistoryBean() {}

 public Collection<HistoryEntry> getItems() {
  SessionRenderer.addCurrentSession("chat");
  return history;
 }

 public void appendEntry(String id, String text) {
  HistoryEntry he = new HistoryEntry();
  he.setUid(id);
  he.setText(text);
  history.add(he);
  if( history.size() > maxSize ) history.remove(0);
  //Trigger ICEFaces
  SessionRenderer.render("chat");
 }
}
```

The code holds a history of the last 15 entries. When the history items are retrieved with the "getItems()" method call, the ICEfaces "SessionRenderer" is added with the label "chat." Then, when a new history entry is appended at the end of the "appendEntry()" function, the "SessionRenderer" is invoked again, this time with a call to render. That in turn would start the asynchronous update event. These are the only two lines of code that need to be added to the Java source code. The rest of the ICEfaces framework is configuration, with some altering in the main JSP page.

```xml
<?xml version="1.0" encoding="UTF-8"?>
<f:subview id="history" xmlns:f="http://java.sun.com/jsf/core"
     xmlns:h="http://java.sun.com/jsf/html">

  <html>
   <head>
    <meta http-equiv="Content-Type" content="text/html; charset=UTF-8"/>
    <title>JSP Page</title>
```

```
  </head>
  <body>
   <h2>Chat History</h2>
   <f:subview id="historytable">
    <h:dataTable id="history" value="#{HistoryBean.items}" var="msg">
     <h:column>
      <f:facet name="header">
       <h:outputText value="Identity"/>
      </f:facet>
      <h:outputText value="#{msg.uid}"/>
     </h:column>
     <h:column>
      <f:facet name="header">
       <h:outputText value="Message"/>
      </f:facet>
      <h:outputText value="#{msg.text}"/>
     </h:column>
    </h:dataTable>
   </f:subview>
  </body>
 </html>
 </f:subview>
```

Here the "history.jspx" file is defined. Note that this is all standard JSF syntax without any additions. The only place where the ICEfaces framework needs to intercept is in the "index.jsp" page.

```
<html>
  <head>
    <meta http-equiv="Content-Type" content="text/html; charset=UTF-8">
    <title>JSP Page</title>
  </head>
  <body>

  <h1>JSP Page</h1>
    <jsp:forward page="history.iface" />
  </body>
</html>
```

Instead of forwarding the main page to "history.jspx," it does a forward to "history.iface." The interception is done by a special ICEfaces Servlet, so we need to look at the "web.xml" file.

```
<?xml version="1.0" encoding="UTF-8"?>
<web-app version="2.5" xmlns="http://java.sun.com/xml/ns/javaee"
xmlns:xsi="http://www.w3.org/2001/XMLSchema-instance" xsi:schema
Location="http://java.sun.com/xml/ns/javaee http://java.sun.com/xml/
ns/javaee/web-app_2_5.xsd">
```

```xml
<context-param>
 <param-name>javax.faces.STATE_SAVING_METHOD</param-name>
 <param-value>server</param-value>
</context-param>
<context-param>
 <param-name>com.sun.faces.validateXml</param-name>
 <param-value>true</param-value>
</context-param>
<context-param>
 <param-name>javax.faces.DEFAULT_SUFFIX</param-name>
 <param-value>.jspx</param-value>
</context-param>

<listener>
 <listener-class>
  com.icesoft.faces.util.event.servlet.ContextEventRepeater
 </listener-class>
</listener>

<servlet>
 <servlet-name>Faces Servlet</servlet-name>
 <servlet-class>javax.faces.webapp.FacesServlet</servlet-class>
 <load-on-startup>1</load-on-startup>
</servlet>

<!-- - Persistent Faces Servlet -->
<servlet>
 <servlet-name>Persistent Faces Servlet</servlet-name>
 <servlet-class>com.icesoft.faces.webapp.xmlhttp.PersistentFacesServlet
 </servlet-class>
 <load-on-startup> 1 </load-on-startup>
</servlet>

<!-- Blocking Servlet -->
<servlet>
 <servlet-name>Blocking Servlet</servlet-name>
 <servlet-class>com.icesoft.faces.webapp.xmlhttp.BlockingServlet
 </servlet-class>
 <load-on-startup> 1 </load-on-startup>
</servlet>

<!-- Persistent Faces Servlet Mappings -->
<servlet-mapping>
 <servlet-name>Persistent Faces Servlet</servlet-name>
 <url-pattern>/xmlhttp/*</url-pattern>
</servlet-mapping>

<servlet-mapping>
 <servlet-name>Persistent Faces Servlet</servlet-name>
 <url-pattern>*.iface</url-pattern>
</servlet-mapping>
```

```
<!- - Blocking Servlet Mapping ->
<servlet-mapping>
 <servlet-name>Blocking Servlet</servlet-name>
 <url-pattern>/block/*</url-pattern>
</servlet-mapping>

<servlet-mapping>
 <servlet-name>Faces Servlet</servlet-name>
 <url-pattern>*.jsf</url-pattern>
</servlet-mapping>

<session-config>
 <session-timeout>30</session-timeout>
</session-config>
<welcome-file-list>
 <welcome-file>index.jsp</welcome-file>
</welcome-file-list>
</web-app>
```

The "web.xml" file is quite large, so if you ever wondered why annotations were introduced in JEE 5, this is your answer. First of all, ICEfaces sets up a couple of context parameters, the most important one of which is "DEFAULT_ SUFFIX," which specifies what we are dealing with in the "jspx" files. This is due to the fact that we mapped the "history.jspx" to "history.iface." The next specific declaration is the "ContextEventListener," which the ICEfaces framework needs to set up. Then there is a standard Faces Servlet declaration, and after it come two ICEfaces specific Servlets. One is for blocking the COMET connection, and the other maps the "*.iface" where our "history.iface" would be intercepted. The rest of the "web.xml" file is standard configuration.

The last configuration file to be modified is the "faces-config.xml":

```
<faces-config xmlns="http://java.sun.com/JSF/Configuration">
 <application>
  <view-handler>
   com.icesoft.faces.facelets.D2DFaceletViewHandler
  </view-handler>
 </application>

 <managed-bean>
  <managed-bean-name>HistoryBean</managed-bean-name>
  <managed-bean-class>net.sipservlet.sample.chatserver.HistoryBean
  </managed-bean-class>
  <managed-bean-scope>application</managed-bean-scope>
 </managed-bean>
</faces-config>
```

At the application scoping level, the "D2DFacletViewHandler" from the ICEfaces framework gets initialized, and we declare the History Bean according to standard JSF, also in application-wide scope.

As well as making the History Bean available in the SIP Servlet scope, we need to add it to the "ServletContext." There is a special listener in "javax.servlet" that gets called when a Servlet context is created (as an effect of deploying the ".sar" file). Now everything is in place, but the "SipChatServlet" code is not described; however, after coming this far in the book, these details easily could be gleaned by studying the sample code.

Now it is time to build the ".sar" file and to deploy it. For testing purposes, we can utilize the X-Lite and the "SipCommunicator" SIP clients. One is set up to host user "sip:chris@sipservlet.net," while the other has "sip:stoffe@sipservlet.net." In both clients, a contact representing the forum "sip:DefaultForum@sipservlet.net" should be added. It will be used to submit messages to the forum.

It is time to open up a browser and access the ICEfaces enabled page.

Note This time the application is deployed at the root, and this is a trick that can be done in the "sun-web.xml" file deployment descriptor by setting "<context-root>/</context-root>."

http://localhost:8080 will open up an empty chat history page.

After sending some messages between Chris and the forum and Stoffe and the forum, the page would look like Figure 9.2.

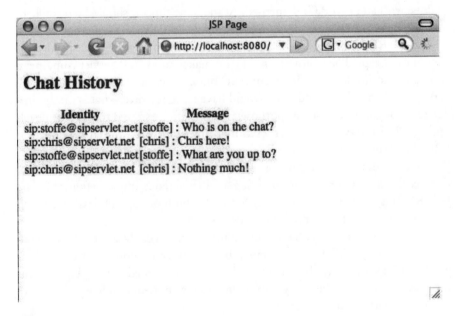

Figure 9.2 Java Server Faces with ICEFaces chat Web GUI.

No, the screenshot does not say much: The interactivity should be experienced live!

This shows the potential of the ICEfaces framework, and what is really nice is that a skilled JSF developer can create a site, while a skilled SIP Servlet developer can add the communications part and the COMET asynchronous updates.

9.4 REST and JAX-RS

Representational State Transfer, commonly abbreviated REST, is an architectural concept that was first presented by Roy Fielding. It describes how to represent objects over HTTP protocol using the standard methods like GET, POST, PUT, and DELETE. The actual object is specified by an HTTP URL. (See Dr. Fielding's dissertation at www.ics.uci.edu/~fielding/pubs/dissertation/top.htm.)

Many services built on top of HTTP apply the REST pattern. One such service that is used in conjunction with SIP is the XCAP protocol, in which the buddy list of a chat and presence application retrieves, stores, updates, and deletes its entries.

To give a realistic example in the context of SIP Servlets, we could imagine that a registrar Servlet state could be made available. Going to the base URL of the resource, we would be able to retrieve a list of all currently registered users and their user agent contacts. Then, by further navigating down per each registered user, it would be possible to retrieve a specific user data object. The data retrieval would be done with an HTTP GET method. At the same time, it should be possible to add a new contact entry for a user by utilizing the HTTP POST method.

Since REST is only a concept, there is need of an implementation for realizing the concept. The history from HTTP makes an HTTP Servlet container a good place to implement such a concept. But even with the support of HTTP Servlets, there is a lot of code that would have to be rewritten that is pretty similar every single time. For this purpose, a new JSR was started to give the developers of REST services a higher level of abstraction. JAX-RS that is defined in JSR 311 is specified to make REST services easier and quicker to implement. When it comes to the JCP process that defines how a JSR should be conducted, there is also a requirement of providing a RI (Reference Implementation). The JAX-RS RI is called Jersey and, lucky for us, it is built on top of GlassFish, which in turn happens to be the on same platform as SailFin.

By taking an ordinary SailFin distribution, we can easily add the Jersey RI to be cohosted on a SailFin instance. First, the SailFin server should be stopped. It is good then to start out fresh with a new domain, as described in Chapter 7. Then, the easiest way to add the Jersey implementation, is to use the "updatecenter" tool that comes with SailFin.

```
sailfin/updatecenter/bin/updatetool
```

A graphical tool (see Figure 9.3) should pop up on the screen.

In the list, the Jersey server should be checked, then press Install. The license GUI pops up and, after accepting the Jersey component, is installed on to the Sail-Fin server instance. Note that the version of Jersey is 0.8, so it is not yet finalized. Its quality is quite good, so it should be no major obstacle. However, the implementation might change, since it is quite a new component, so for this reason the update center also provides an upgrade functionality. After Jersey is installed, it should appear under the Installed Software tab (see Figure 9.4). Here we are able to see whether an update is released, and also we would be able to get the update.

This is all that needs to be done to install JSR 311 JAX-RS support. Under the SailFin directory, there should now be a Jersey folder. The folder contains the necessary libraries but also the documentation and some samples. Now the SailFin server can be started again with the "asadmin" "start-domain" command. The next step is to develop the registration SIP Servlet that we want to expose with REST.

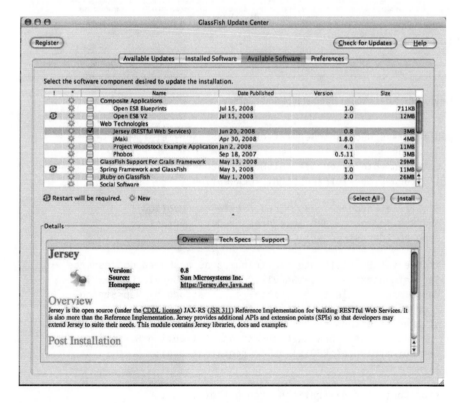

Figure 9.3 Update Center choosing Jersey component to install.

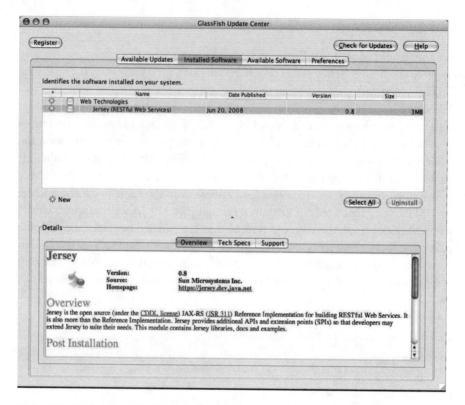

Figure 9.4 After successful installation of Jersey.

```java
public class RegistrarServlet extends SipServlet {
@Override
 protected void doRegister(SipServletRequest req)
       throws ServletException, IOException {

  SipServletResponse resp = req.createResponse(200);
  Address contact = req.getAddressHeader("Contact");
  resp.setAddressHeader("Contact", contact);

  // Use the ServletContext attributes as a shared db resource
  ServletContext sc = getServletConfig().getServletContext();
  Map<String, Collection<String> registrations = (Map<String,
    Collection<String>) sc.getAttribute("REGISTERED");
  if (registrations == null) {
   registrations = new HashMap<String, Collection<String>();
   sc.setAttribute("REGISTERED", registrations);
  }

  storeContact(registrations, req.getTo(), contact);
  resp.send();
}
```

```
public static void storeContact(Map<String, Collection<String>
   registrations, Address to, Address contact) {
  Collection<String> contacts =
registrations.get(to.getURI().toString());
   if (contacts == null) {
    contacts = new ArrayList<String>();
   }
   contacts.add(contact.getURI().toString());

   //Use only URI form the from header as key
   registrations.put(to.getURI().toString(), contacts);
 }
}
```

The "doRegister()" method is quite straightforward. It looks for the SIP "Contact" header; it creates a SIP 200 response and appends the "Contact" header to it. Then the Servlet stores away the "Contact" header into a collection, which you in turn save as an attribute in "ServletContext." This is probably the easiest way to store away things; since the REST framework is built on top of HTTP Servlets, it is quite easy to access the registration state in the Servlet Context.

Note The registrar Servlet is by far not complete. A real one should check more on the SIP "Expires" header, and the 200 should contain the full list of contacts registered for a SIP user. It is a good idea to persist registration information. A common pattern that is often used is to utilize EJB 3 and the Entity Manager for storing the data in a database. To minimize the code and show the REST concept, the Registrar Servlet is kept to a minimum.

```
@Path("/")
public class RegistrarResource {

 @Context ServletContext sc;

 public RegistrarResource() {}

 @GET
 @Produces("application/xml")
 public String getRegistered() {
  Map<String, Collection<String> registered = (Map<String,
    Collection<String>) sc.getAttribute("REGISTERED");
  StringBuilder output = new StringBuilder();
  output.append("<registrations>\r\n");
  for (String id : registered.keySet()) {
   appendUser(output,id,registered.get(id));
  }
  output.append("</registrations>\r\n");
  return output.toString();
 }
```

```
@GET
@Produces("application/xml")
@Path("users/{userid}/")
public String getContact(@PathParam("userid") String id) throws
  ServletParseException {
 Map<String, Collection<String> registered = (Map<String,
   Collection<String>) sc.getAttribute("REGISTERED");
 StringBuilder output = new StringBuilder();
 output.append("<registrations>\r\n");
 appendUser(output,id,registered.get(id));
 output.append("</registrations>\r\n");
 return output.toString();
}


@POST
@Consumes("application/x-www-form-urlencoded")
@Path("users/{userid}/")
public void postContact(@PathParam("userid") String id, String form)
throws
  ServletParseException, IOException {

 Hashtable table = HttpUtils.parseQueryString(form);
 String contact = ((String[]) table.get("contact"))[0];
 Map<String, Collection<String> registered = (Map<String,
   Collection<String>) sc.getAttribute("REGISTERED");
 SipFactory sf = (SipFactory) sc.getAttribute(SipServlet.SIP_FACTORY);
 RegistrarServlet.storeContact(registered, sf.createAddress(id),
   sf.createAddress(contact));
}

private void appendUser(StringBuilder sb, String uid,
Collection<String>
  contacts) {
 sb.append("<user id=\"");
 sb.append(uid);
 sb.append("\">\r\n");
 for (String contact : contacts) {
  sb.append("<contact>");
  sb.append(contact);
  sb.append("</contact>\r\n");
 }
 sb.append("</user>\r\n");
}
}
```

The @Path annotation states that this is a root JAX-RS resource. It is mandatory for there to be one top path resource when using JAX-RS. The next interesting line is the @Context injections. JAX-RS allows for the injection of the "Servlet

Context" that we are going to use for retrieving the saved registration information by the "RegistrarServlet."

For registering what method should be triggered when various HTTP calls are made, each resource function needs to be annotated with a proper HTTP method annotation. The first method in our resource "getRegistered() "is annotated by a @GET annotation. It does not contain any @Path annotation, so it will be used as the root. The method is also annotated with @Produces. In this example, the returned MIME will be XML formatted. The actual XML schema is made up in order to be easily read in a browser. The code inside the function "getRegistered" simply gets the registered collection and formats a XML document with the help of the "appendUser()," which is iteratively called for every user registered in our registrar. The produced XML string will be appended to the 200 "HttpServletResponse" that Jersey will generate for the incoming HTTP GET.

Looking at the second method, "getContact()," notice that the @GET and @Produces annotations are the same. The difference from the "get Registered()" method is the @Path annotation. It defines a path of users, but the element after the slash shows how dynamic path parameters are declared in the JAX-RS framework. In brackets {userid}, the dynamic "userid" parameter gets declared. Because it is dynamic, the Servlet code knows what it is set to. For this purpose, there is a @PathParameter defined in JAX-RS. In this example, the framework is injecting it to the method input parameter ID by calling:

```
(@PathParam("userid") String id
```

Soon, when we look at a real executed scenario, it will be clear, but first, a short example illustrating what is happening:

Accessing pathroot"/" would yield a complete list of all registered resources. A call to "/users/stoffe" would inject the String "stoffe" into the ID parameter, while "/users/chris" would inject "chris" into the ID parameter. The rest of the method "getContact()" does the same thing as the "getRegistered()," except it only produces XML output for one user.

The third JAX-RS resource method is quite different. It is annotated with @POST, and it consumes an HTTP form mime type (@Consumes). Since it consumes a MIME type object, it is intended to be used for inserting or updating an entry rather than for getting operations like the previous examples methods demonstrated. The "postContact()" method also utilizes the dynamic path parameter syntax as the "getContacts()" does. It will also get a string that represents the form string. "HttpUtils" is used to parse the contact parameter that is submitted in the form. Then the registration collection is searched to find the user specified by the dynamic path parameter ID. When the right user is found, the contact is then appended so that the user's available contacts are returned.

This is all that needs to be done codewise. The next step is to look at the "web.xml" file.

```xml
<?xml version="1.0" encoding="UTF-8"?>
<web-app version="2.5" xmlns="http://java.sun.com/xml/ns/javaee"
xmlns:xsi="http://www.w3.org/2001/XMLSchema-instance" xsi:schemaLoca-
tion="http://java.sun.com/xml/ns/javaee
http://java.sun.com/xml/ns/javaee/web-app_2_5.xsd">
  <servlet>
    <servlet-name>ServletAdaptor</servlet-name>
    <servlet-class> com.sun.jersey.spi.container.servlet.Servlet
Container
    </servlet-class>
    <load-on-startup>1</load-on-startup>
  </servlet>
  <servlet-mapping>
    <servlet-name>ServletAdaptor</servlet-name>
    <url-pattern>/resources/*</url-pattern>
  </servlet-mapping>
  <session-config>
    <session-timeout>
      30
    </session-timeout>
  </session-config>
  <welcome-file-list>
    <welcome-file>index.jsp</welcome-file>
  </welcome-file-list>
</web-app>
```

The JAX-RS is hooked into the application by defining the "ServletContainer" Jersey Servlet. It is named to "ServletAdaptor," and it is mapped in the converged application under "/resources."

Note It is easiest to use NetBeans to produce both the SIP Servlet code and "sip.xml" file. Then REST support can be added, generating the Resource class and "web.xml" file together with inclusion of the needed JAX-RS libraries.

Right-click on the Converged SIP Servlet project, and choose "RESTful Web Services from Patterns" (see Figures 9.5 and 9.6).

Pick the singleton pattern.

Name the resource and path and click finish (see Figure 9.7).

The next step is to build the application and then deploy it. We will need to register some clients in one of two possible ways. The SIP Test Agent from Net-Beans could generate the SIP registration. Another alternative is to use real SIP clients. The screenshot in Figure 9.8 is an X-Lite client that was used to register Chris and a "SipCommunicator" to register Stoffe. Browsing to the root URL of

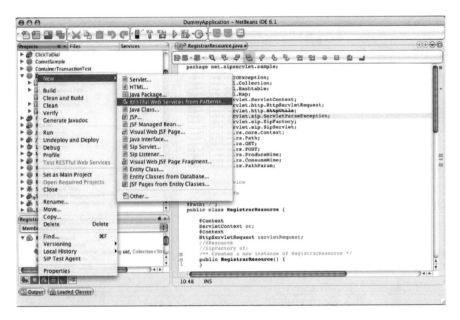

Figure 9.5 NetBeans Wizard creating a REST Web service.

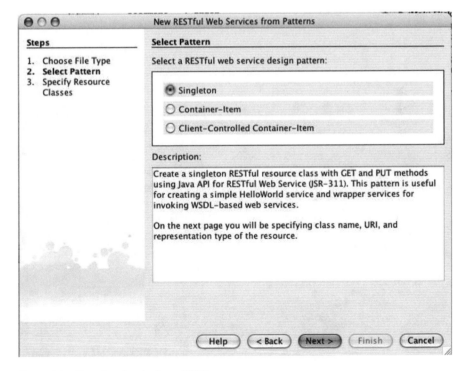

Figure 9.6 Choosing the singleton REST pattern.

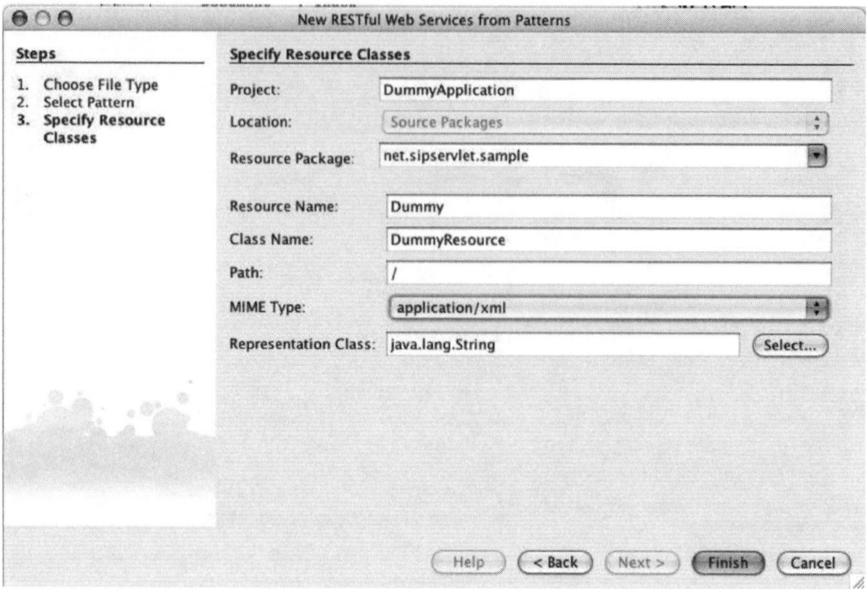

Figure 9.7   Naming the REST Resource and setting its path.

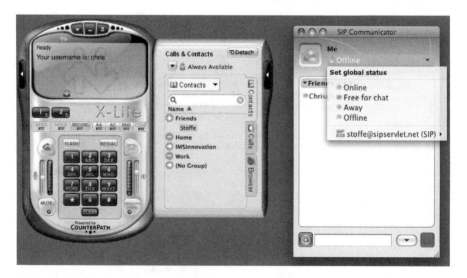

Figure 9.8   Chris on the X-Lite client and Stoffe on the SIPCommunicator client.

the application yields a JSP page http://loaclhost:8080/RegistrarRestServer. This is done so that a standard Web application can be deployed. To reach the root path of the REST resources, append "/resources" to the URL (according to the mapping from "web.xml").

Figure 9.9 is how the registrations from X-Lite and "SipCommunicator" would look like.

Now, looking at a particular user, "/users/sip:chris@sipservlet.net is appended" (see Figure 9.10).

We could also look at "/users/sip:stoffe@sipservlet.net/," which would result in the other registration for user Stoffe. This shows the @GET annotations of resources at work. In order to test the @POST annotation, we could use the command line UNIX utility curl.

```
curl -d contact=sip:voicemail@sipservlet.net
http://localhost:8080/RegistrarRestServer/resources/users/sip:stoffe@
sipservlet.net/
```

The result of running curl would look like Figure 9.11 in the browser.

Under the user Stoffe, the new voicemail contact has been appended. This example showed the basics, but a more common usage of JAX-RS is to produce JSON-formatted documents. The benefit of formatting the MIME in JSON is that it produces compliant JavaScript code. Then the registrar service can be easily mashed up by HTML developers. There is an excellent example of how to produce

Figure 9.9 Top level listing all REST registration resources.

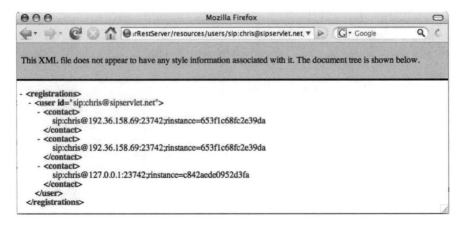

Figure 9.10   Listing only "Chris" registration by specifying user's path.

JSON-formatted documents installed under "sailfin/jersey/examples/JsonFrom Jaxb": As the name implies, JAXB is used to make a Java object to XML mapping, and then the "com.sun.jersey.api.json.JSONJAXBContext" class allows for serialization and deserialization of JAXB beans into the JSON format and vice versa. More resource and discussions can also be found at https://jersey.dev.java.net/.

In order to make the created REST services easier to consume, there is a description language that services can utilize, called Web Application Description Language (WADL), a standardization attempt to format the service description so that development tools can be utilized to autogenerate code stubs for various different languages and frameworks. Since the service is in the end an HTTP serv-

Figure 9.11   Listing after running "curl." User "Stoffe" has a new contact field.

ice, it does not matter if one side is written in Java while the other uses Perl or some other framework. The idea behind it is to make a counterpart to Web Services (JAX-WS) and the Web Services Description Language (WSDL). WS is perceived by some as very complicated and also, in some cases, too powerful and too hard to understand for reaching a broad industry market. This is by no means an attempt to favor one or the other here. Many HTTP-based Web 2.0 services have managed to become very popular in "mashups," and by describing them in WADL, the chances of getting them used in even more "mashups" increases.

Note   There is an interesting project at https://wadl.dev.java.net/ that provides a "wadl2java" tool. There are also links on the site to the WADL specifications.

9.4.1   Consuming a REST Service

Now that we have produced a REST service, we might also be interested in a clean way of consuming a REST service. When packaged in a nice way, a REST service should be easily consumed. In NetBeans, there is a Services tab, where some REST and WS services are collected. To continue the spin on the registrar application, a nice, simple showcase would be to mash it up with the Twitter service.

The first step is to expand the services tag under the Twitter service (see Figure 9.12).

For this sample, the "updateStatus" method would be implemented. It is also possible to view the entire Twitter WADL by right-clicking on "What Are You Doing Service" (see Figure 9.13).

By selecting "updateStatus" and dragging it to the place in the SIP Servlet code, NetBeans will be triggered to generate code stubs and include the needed libraries into the project. The generated code for the Twitter "updateStatus" method is as follows:

```
try {
  String status = "";
  String format = "xml";

  RestResponse result = TwitterWhatAreYouDoingService.updateStatus
(status,format);

  twitter.whatareyoudoingservice.twitterresponse.StatusType resultObj =
    result.getDataAsObject(
    twitter.whatareyoudoingservice.twitterresponse.StatusType.class );

  //TODO - Uncomment the print Statement below to print result.
  //System.out.println("The SaasService returned: "+result.getDataAs
String());
} catch (Exception ex) {
  ex.printStackTrace();
}
```

Figure 9.12 Choosing Twitter "updateStatus" function to auto generate code stub.

Figure 9.13 Right click to view the WADL definition for the Twitter service.

So after slight restructuring and adding it to the registrar servlet code the result looks like this:

```
@javax.servlet.sip.annotation.SipServlet
public class RegistrarServlet extends SipServlet {

 protected void doRegister(SipServletRequest req)
   throws ServletException, IOException {

  SipServletResponse resp = req.createResponse(200);
  Address contact = req.getAddressHeader("Contact");

  resp.setAddressHeader("Contact", contact);
  setTwitterStatus("Could now be contacted on : " + contact.getURI());
  resp.send();
 }

 private void setTwitterStatus(String status) {
  try {
   RestResponse result = TwitterWhatAreYouDoingService.updateStatus
     (status,"xml");

   StatusType resultObj = result.getDataAsObject(
       twitter.whatareyoudoingservice.twitterresponse.Status
Type.class);

   log("The SaasService returned: " + result.getDataAsString());
  } catch (Exception ex) {
   log("Failed update Twitter",ex);
  }
 }
}
```

The code is now complete, but there is one more thing that needs to be done. For executing the update command, we need to specify the Twitter account that is going to be updated. Not only that, but the "updateStatus" also requires authentication information, so we also need the password for the corresponding Twitter account (see Figure 9.14).

As can been seen in Figure 9.14, we can include in the inserted classes using NetBeans. They are prefixed by "org.netbeans.saas" and "org.netbeans.saas. twitter." In the second of these packages is a properties file that requires the programmer to enter the user name and password. When this is done and all files are saved, it is time to compile the ".sar" file. Then we need to deploy it and make a SIP registration. To demonstate this, we can once more register Stoffe with the help of the "SipCommunicator" client. Now, going to www.twitter.com and logging in to the same account where we entered in the properties, we can see the result in Figure 9.15.

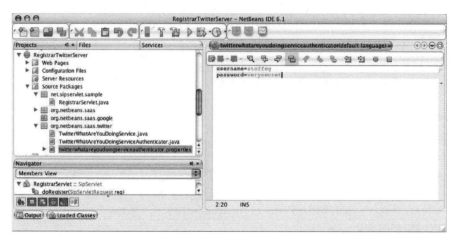

Figure 9.14   Specifying developer specific properties for the Twitter service.

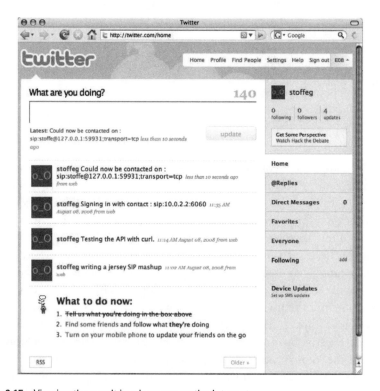

Figure 9.15   Viewing the result in a browser on the Internet.

That concludes this simple sample. One thing that is worth pointing out is that the scenario is probably not the best. It was chosen for its simplicity and alignment with the previous ones. Instead, it would make more sense to use the SIP PUBLISH method that reports presence updates for a user. This in turn involves parsing presence XML documents, and there are various formats of the XML as well. Another use case could be to update the Twitter status on an incoming call, and then on the BYE SIP message, update again when the user is off the call.

There is also a client-side REST implementation in the Jersey classes. It tends to be used internally for testing services, but it could also be exploited to do server-side "mashups" in a ".sar" or ".war" application.

9.5 Java ME JSR 180

JSR 180 is a specification defined for Java Micro Edition (Java ME). Java ME has two different base configurations, Connected Device Configuration (CDC) and the Connected Limited Device Configuration (CLDC 1.1—JSR 139). Most smart phones, IP gateways, and TV setup boxes implement the CDC configuration. The majority of the mobile phones in the market support the CLDC configuration with the Mobile Information Device Profile (MIDP 2.x—JSR 118).

Java ME programming is an altogether different kind of paradigm. There are basically two options. One is to use a phone that already has the JSR 180 support built in. The other alternative is to use a library that implements the SIP support using the raw socket support in MIDP. The support for JSR 180 was increased in deployment since the umbrella JSR 248 Mobile Service Architecture (MSA) promotes its implementation. The fact that mobile phone vendors want to be MSA compliant has resulted in many devices on the market supporting SIP in their Java ME environment (see Figure 9.16).

As is often the case with Java ME JSRs, the API can behave a little differently between the different phone vendors. At the same time, it is not too uncommon for the real mobile phone and the emulated development environment to have slightly different behaviors. To complicate things further, there can be two modes of operations of the JSR 180 stack. The specification talks about a dedicated mode

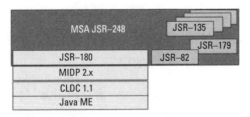

Figure 9.16 JSR 180 as part of Mobile Service Architecture in the Java ME API stack.

in which the Midlet (Mobile applet and code entry point) is the only client using the underlying SIP stack. The other mode is the shared mode, in which the SIP stack is shared among other Midlets and native phone applications. Usually, the programmer is responsible for opening the stack and configuring it while in the dedicated mode. There will be no clashes as long as there is no other Midlet or native application using the same socket port for incoming signaling. If you choose port 5060 for incoming SIP traffic, there might be a possible clash, and thus the connection would fail.

When a shared mode stack is used, normally the phone comes with a configuration GUI, in which SIP connectivity is configured. It is then up to the phone software to send SIP REGISTER messages and initialize the stack before any Midlet opens up the connection.

For the people who are new to Java ME MIDP programming, the main class, and the entry point to your software, is a class that extends "javax.microedition.midlet.Midlet."

The initialization of the code normally happens in the "startApp() callback."

```
Public class ChatMidlet extends Midlet {
      Public void startApp {
            //Init the SipConnection
            //Init and render the GUI components
      }
}
```

In normal cases, the GUI components would render a text input box, a destination SIP address text box, and a send button in just the same way as one creates the client using HTML.

Then, pushing the send button produces a SIP message to be sent.

The initialization of the SipConnecion Notifier can be done like the following sample code:

```
SipConnectionNotifier scn;

//Dedicated Mode init
scn = Connector.open("sip:5060;transport=tcp");

// Shared Mode init
scn = Connector.open("sip:*;type=text/chat-sample");
```

In the dedicated mode, initialization of the client will act as a UAS on TCP port 5060. All the received messages arriving on that port will be delivered to the "SipConnectionNotifier."

While in the shared mode the socket is predefined by the phone. For that reason no port information is specified but instead a type = text/chat-sample.

First of all the phone needs to be registered in order for the stack to operate. Then all messages that contain a SIP request header of:

```
Accept-Contact: *;type=application/chat-sample
```

This will cause an incoming SIP message to be routed to the very same instance that was initialized in the shared mode open command. Of course, two applications cannot register the same type pattern, or else the second one opening will fail. Basically, this is branching on a particular port or "Accept-Contact" header.

Note For the sample, Sun Wireless Toolkit 2.5.2, which uses the shared mode, was used, but for real execution a Sony Ericsson JP-8 (W890), which supports only the dedicated mode, was used. So, a developer really needs to test depending on the intended real platform to verify against.

The "ChatMidlet" class also defines some basic MIDP classes for input and output, but it is to be considered more as a minimalistic GUI approach. A lot of time can be spent on designing a proper commercial-grade GUI.

```
public class ChatMidlet extends MIDlet implements CommandListener {
  Display d = null;
  Form f = new Form("ChatForumClient");
  Command messageCmd = new Command("Send", Command.ITEM, 1);
  Command exitCmd = new Command("Exit", Command.EXIT, 1);
  TextField messageField = new TextField("Message to send", "", 40,
  TextField.ANY);
  TextField addressField = new TextField("Target IP",
  "sipservlet.net", 40, TextField.ANY);
  TextField userField = new TextField("User", "bob", 40,
  TextField.ANY);
  TextField portField = new TextField("Port", "5060", 5,
  TextField.DECIMAL);
  TextField chatForumField = new TextField("Chat forum to send",
  "DefaultForum", 40, TextField.ANY);
```

The Display class is the main entry point for the graphic screen. The Form is a container component in which other items can be added. In this example, only two controls are used. The first is a Command that will be tied to menu buttons to interact with the user. The other is the "TextField," in which a Midlet user is able to read and update various fields. The class declaration also contains a "CommandListener" interface that specifies the "commandAction" callback function to be called when the user presses one of the command buttons.

One of the Midlet class callbacks is the "startApp" function, by which all the commands and text fields would be initialized.

```
protected void startApp() throws MIDletStateChangeException {
 d = Display.getDisplay(this);
 f.addCommand(messageCmd);
 f.addCommand(exitCmd);
 f.append(messageField);
 f.append(chatForumField);
 f.append(addressField);
 f.append(userField);
 f.append(portField);
 f.setCommandListener(this);
 d.setCurrent(f);
 try {
  initConnectionNotifier();
  ReceiverThread rt = new ReceiverThread();
   new Thread(rt).start();
   } catch (IOException e) {
   e.printStackTrace();
  }
}
```

The commands and text fields are added to the form, and then the form is set in the Display class. This is a quite standard Java ME programming pattern. In a larger application, normally there would be multiple forms, and only the one currently displayed would be set. After the graphics initialization, the connection notifier is also called. This function sets up the shared or dedicated mode "Sip ConnectionNotifier." Next is the creation of a new thread called "ReceiverThread." Since SIP is asynchronous, SIP messages can arrive independently. For this purpose, the Midlets has a thread that constantly monitors for new incoming SIP messages. There is an alternative to this design pattern. The "SipConnectionNotifier" also has a callback interface that can be called when a new message has arrived.

In this sample, we use two thread design patterns. One thread is receiving the messages while the other thread is sending new SIP messages. It is more related to how the application should behave. If messages are sent and received in the same thread that graphics are, then it is not possible to navigate when dealing with a SIP message. This can be either good or bad depending on how the developer wants the GUI to behave.

Let's look at the "CommandListener" callback function to see what happens when a message is sent.

```
public void commandAction(Command cmd, Displayable disp) {
  if (cmd == messageCmd) {
   sendSipMessage(prot + ":" + chatForumField.getString() + "@'+
    addressField.getString() + ";transport=" + transport,
    messageField.getString());
  } else if (cmd == exitCmd) {
   try {
```

```
    destroyApp(false);
    notifyDestroyed();
  } catch (MIDletStateChangeException ignore) {}
  }
}
```

The first command is issued when the Send button is invoked. That calls the "sendSipMessage" function. The input to the function is the target SIP URI in string format. In this case, when defaults are used, it will produce "sip:Default Forum@sipservlet.net;transport=udp." The second argument is also a string, and it will have to be entered in the "messageField" from the GUI. The third command is the exit command that is present in most Java ME applications. It is good practice to let the user exit a Java ME application from the root menu. On some phones it might actually be quite hard to find another way of killing a running application.

Look at the code sending the SIP message to the chat forum:

```java
public void sendSipMessage(String uri, String msg) {
 SenderThread st = new SenderThread(uri, msg);
 new Thread(st).start();
}

class SenderThread implements Runnable {
 String uri;
 String encoding;
 String contentType = "text/plain";
 byte[] content;

 // Constructor for text messages
 public SenderThread(String uri, String msg) {
  this.uri = uri;
  if (msg != null) {
   encoding = "UTF-8";
   try {
    content = msg.getBytes(encoding);
   } catch (UnsupportedEncodingException ignore) {}
  }
 }

 public void run() {
  SipClientConnection sc = null;
  try {
   sc = (SipClientConnection) Connector.open(uri);
   sc.initRequest("MESSAGE", scn);
   sc.setHeader("From", prot + ":" + userField.getString() + "@" +
addressField.getString());
    if (content != null) {
     sc.setHeader("Content-Type", contentType);
     sc.setHeader("Content-Length", Integer.toString(content.length));
```

```
        sc.setHeader("Contact", "sip:" + userField.getString() + "@" +
scn.getLocalAddress() + ":"
          + scn.getLocalPort() + ";transport=" + transport);
        if (encoding != null) sc.setHeader("Character-Encoding", encoding);
        OutputStream os = sc.openContentOutputStream();
        os.write(content);
        os.close(); // close stream and send the message
      }
      // wait maximum 35 seconds for response
      boolean ok = sc.receive(35000);

      sc.close();
      pauseApp();
    } catch (Exception ex) {
      ex.printStackTrace();
    }
  }
}
```

From the source code, we can see that the "sendSipMessage" function simply creates a new "SenderThread" and starts it. The constructor of the thread stores away the target SIP URI and the message to be sent. In the "run" function triggered by the start of the thread, first a "SipClientConnection" is opened toward the target URI. The call to "initRequest" specifies that it is a SIP MESSAGE method we are going to use. Then, based on the content, both the "Content-Length" and "Content-Type" SIP headers are added. The next header added is the SIP "Contact" header, and this is a little bit of a hack, since normally the SIP MESSAGE would not contain such header. This is only to save us some programming; in normal operations, responses should be sent on the "From" header in the chat server. In order to minimize, we never did send any SIP register, so that lookup would fail; thus, we provide the path back in the "Contact" header (this could be any header name, really, since we will also write the chat server part). The last header specifies the character encoding, and it uses the UTF-8 encoding from the thread constructor.

Some basic headers, such as "To," "CSeq," "Call-ID," and "Via" are set by the framework together with the Request URI. When all the headers are in place, the body needs to be appended. For this reason, an "OutputStream" can be fetched from the "SipClientConnection." Calling the "write" function with the content for the body will append it to the message. The next call is to the "Output Stream" function "close." This will actually flush the stream and send away the SIP message. When this is done, we then call the "receive" function on the client connection and also specify the amount of time it should block. In this example, we specify 35 seconds: A SIP transaction is normally T1*64, which is 32 seconds. This should give us plenty of time to receive a final response. The return type of the "receive" function is a Boolean, but we do not care to do anything with the result.

If we would receive a 2xx success SIP response, then the Boolean would evaluate to "true."

Now that we have shown how to send a message, let's continue with the code for receiving a message. As was shown previously, a "ReceiverThread" has already been started in the "startApp" Midlet function.

```
class ReceiverThread implements Runnable {
 public void run() {
  while (true) {
   try {
    SipServerConnection conn = scn.acceptAndOpen();

    String cl = conn.getHeader("Content-Length");
    String ct = conn.getHeader("Content-Type");
    int len = Integer.parseInt(cl);
    InputStream is = conn.openContentInputStream();
    byte[] msg = new byte[len];

    int offset = 0;
    int bytesRead = 0;
    while ((bytesRead = is.read(msg, offset, len - offset)) != -1) {
     offset += bytesRead;
     if (offset >= len) break;
    }

    if (ct.startsWith("text/")) {
     f.append(new String(msg) + "\n");
     Alert alert = new Alert("Message");
     alert.setString("Received : " + new String(msg));
     alert.setTimeout(5000); // 5 seconds
     d.setCurrent(alert, d.getCurrent());
    }
    is.close();

    conn.initResponse(200);
    conn.send();
    conn.close();
   } catch (SipException e) {
    e.printStackTrace();
   } catch (IOException e) {
    e.printStackTrace();
   } catch (Throwable t) {
    t.printStackTrace();
   }
  }
 }
}
```

This thread is implemented in a busy loop that would call the "acceptAnd Open" function on the "SipConnectionNotifier." It would block on it until a SIP message arrives. When a message is available, we would then get a handle on it in the form of a "SipServerConnection" object. On the incoming message, we can read any SIP header, and in this case, we are interested the "Content–Length" and the "Content–Type" SIP headers. Knowing the length, then we can read the body of the chat forum message. Since SIP messages can contain different content types, we also make sure that we have got a text message before we display it.

Note The X-Lite SIP client, to name one such client, is sending the application/imis-composing+xml MIME SIP message body content. The messages are sent to indicate that the other part in the communication is in the process of typing. It is better to filter out these messages if they are not significant, but they could also be graphically indicated to the end user.

If the message is text MIME, then we would use two different techniques to show it to the user. First, we append it to the main form of the display. The next lines will show how to use the Alert component in Java ME. It will temporarily take over the display, showing the received message. After a configurable time-out, the control will be handed back to the main form. In this case, the message is displayed in 5 seconds. If other messages arrive within the 5 seconds, they will only be appended to the form, because the Alert component still has the main control. But, after the main form gets the control back, all messages are then visible. Maybe it is not the nicest chat forum GUI, but it should simply demonstrate the basics.

After the message is processed, it is time to send a 200 OK SIP message back. On the "SipServerConnection" we call "initResponse(200)" and then call "send." A more robust client would also send error responses back. We close the connection, and then we go back in the busy loop to accept the one next in line.

This is all the code that has to be written on the client side. Then a Java ME Midlet is usually stored in a ".jar" file. In the archive, the class files would be included, as well as a Manifest file with some basic configuration. The same entries as in the Manifest are usually copied to a stand-alone file called ".jad." The Java Archive Description (JAD) file is usually shipped together with the corresponding ".jar" file, which is used for installing it on a mobile phone. Except for the configuration entries, it is a good practice to sign the ".jad" file with a certificate. This in turn makes the application more trusted in the mobile phone, which is a good idea, since trusted applications can be configured so that security question does not pop up all the time while running a Midlet.

```
MIDlet-1: ChatClient,,net.sipservlet.sample.mechat.ChatMidlet
MIDlet-Jar-URL: MeChat.jar
MicroEdition-Configuration: CLDC-1.1
MIDlet-Version: 1.0.0
MIDlet-Name: ME SIP Chat
```

```
MIDlet-Description: Sample jsr180 and SailFin on server side
MIDlet-Vendor: Ericsson
MicroEdition-Profile: MIDP-2.0
MIDlet-Permissions: javax.microedition.io.Connector.sip
```

Here the name of the main Midlet is specified together with the various versions of the API used. The name of the corresponding ".jar" file is stated with a textual description. The last line indicates that this Midlet is using the SIP JSR 180 support. Signing it is enough to indicate that this application is trusted to send SIP messages; then the Midlet user will not be queried anymore.

To build the ".jar" and ".jad," it is quite convenient to use a development environment. Both NetBeans and Eclipse have special Java ME plug-ins in which the path to the Wireless Toolkit can be specified. In the projects, a reference to a certificates file can be pointed out. When developing and testing the application in the WTK, it might be a good idea to reconfigure it slightly. One thing that is convenient is to disable the security checks. This can be done by changing the default level of security in the WTK configuration. The other convenient thing is to enable network monitoring. The monitoring will start a separate GUI showing all the SIP messages going in and out to the emulator.

When the code is in place and the ".jar" and ".jad" files are produced, we can run the sample. We will reuse the SIP chat server that we used for the ICEfaces example, but first we need to do two adoptions. The first is due to the fact that we have a shared mode SIP stack in the WTK. The outgoing SIP messages from the chat should contain the "Accept SIP" header.

```
req.setHeader("Accept-Contact", "*;type=\"application/chat-sample\"");
```

This will be set on all the "SipServletRequest" instances going out of the "SipContainer." The "Contact" header patch described earlier is shown below. This is an optimization that does not require a SIP re-registration. The "Contact" header is not mandatory in a SIP MESSAGE but if it is present, the server can use the header as the remote target to which to send messages back.

```
String contact = req.getHeader("Contact");
if( contact != null && contact.length() > 0 ) remote = contact;
```

Let's start the modified chat server and try once more with Chris on the X-Lite, Stoffe on the "SipCommunicator," and now Bob on a Java ME client!

Figure 9.17 is the view of the WTK emulator where the main form can be seen with the five text fields. In the bottom, the two Command actions have been put on two of the phone's action buttons. In the lower part of the screen, the "f.append()" result of incoming messages can be seen in the lower part of the phone screen. One such printout is the: "(stoffe) what's up this weekend?"

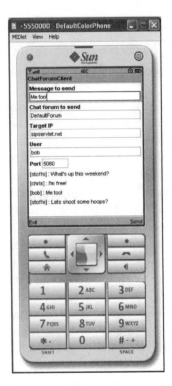

Figure 9.17 WTK Java ME test client running chat application.

Here is the view we saw previously in the ICEfaces COMET-based GUI. The only difference is that we can see Bob sending some messages from his mobile (see Figure 9.18).

And Figure 9.19 is Chris's view as he sees it on the X-Lite SIP client.

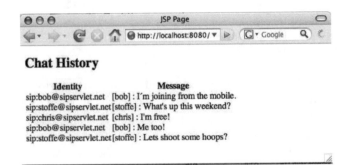

Figure 9.18 Browser-based chat client.

Figure 9.19   X-Lite client chat window.

Reference

[1]   REST introduction on Wikipedia, http://en.wikipedia.org/wiki/REST.

10

The SIP Servlet Application Programming Interface (API)

The topic of SIP Servlet containers and SIP Servlet–based applications has been discussed at great length in this book. The delimitation of roles and responsibilities in the SIP Servlet architecture has become quite clear, with applications being hosted on a compliant container. Some of the major principles of the SIP Servlet architecture are adopted from JEE in general and, more specifically, the HTTP Servlet specification from which SIP Servlets were derived. The ability to create applications that are not only portable across different deployment environments but also portable across differing vendors who comply with the specification is a key objective. At the core of this principle, as with the HTTP Servlet architecture, is a common Application Programming Interface (API). The commonality provided by a containers compliant SIP Servlet API implementation provides the glue that integrates any application with a container. Figure 10.1 provides a high-level illustration of the API layer's role.

Some of the major constructs of the SIP Servlet API have been covered already in other sections of the book, and this chapter aims to be a more complete reference to the workings of the API. Some of the API parts have been left out for clarity, so the reader is encouraged to take a good look at the official API documentation that is provided with the SIP Servlet Specification. The remaining sections of this chapter will provide a high-level guide to primary SIP Servlet API interfaces and their role in successful application creation and deployment.

The main programming interface is defined in the Java package "javax.servlet .sip," which uses the base Servlet 2.5 API as its foundation. The main interfaces have been grouped appropriately to provide more contextual information in relation to roles within the SIP Servlet architecture.

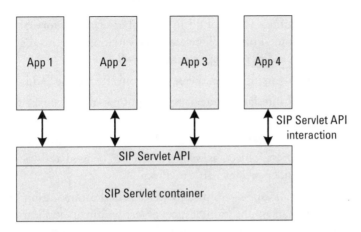

Figure 10.1   SIP Servlet API role.

10.1   Container Utilities

A SIP Servlet container provides a number of utilities that abstract certain complexities away from applications and also allow for integration into larger applications such as JEE. The following are important container-level interfaces that provide such utilities. Some have already been discussed in other sections of the book, which should be referred to if more detail is required.

10.1.1   SipFactory

The "SipFactory" interface can be considered one of the most important in the entire API and provides an application developer with the ability to create a number of other main interface instances (hence "factory"). This includes objects such as "Address," "ApplicationSession," "AuthInfo," "Parameterable," "SipServlet Request," "SipURI," and "URI." All of these objects are covered in the remaining sections of this chapter. A compliant container must make an instance of the SIP Factory available to applications through a "ServletContext" attribute called "javax.servlet.sip.SipFactory." An instance of a SIP Factory can also be injected into an application using the @Resource Java annotation from either within a SIP Servlet or as part of a larger JEE application. More detail on the SIP Factory and on how to utilize it in this way is provided in Chapter 2.

The SIP Factory has the following methods that are used for the creation of other objects used extensively within the SIP Servlet API:

> *createAddress*—"SipFactory.createAddress" method is used to create an "Address" interface instance, which will be covered later in this chapter. An "Address" instance is a representation of an abstracted SIP protocol address

that has a common form found in a number of mandatory SIP headers, such as "To" and "From." There are three variations on this method, each of which takes a different input parameter:

- A string as an input parameter to be converted to the "Address" object;
- A "URI" interface instance as an input parameter to be converted to the "Address" object;
- A "URI" interface instance as an input parameter to be converted to the "Address" object, as well as a string to be used as display name in a SIP header such as "To" or "From."

createApplicationSession—"SipFactory.createApplicationSession" method is used to create an Application Session object within an application. Application Sessions represent an instance of the application and are used to associate any number of SIP signaling interactions with users as well as to store application data. SIP Application Sessions were covered in detail in Chapter 3, and the specific API detail is covered later in this chapter. A variant on this method, called "createApplicationSessionByKey," was introduced in the latest version of the SIP Servlet architecture (Version 1.1) and enables applications to create an Application Session that is associated with a specific, unique key. This enables ease of management and lookup at a later stage.

createAuthInfo—"SipFactory.createAuthInfo" method is used to create an "AuthInfo" interface instance. An "AuthInfo" instance is a convenience object provided by the container for associating security properties related to SIP messaging exchanges such as SIP digests, as covered in the core SIP specification [1]. The "AuthInfo" object is covered later in this chapter.

createParameterable—"SipFactory.createParameterable" method is used to create a "Parameterable" object; this is covered later in this chapter. A number of common SIP headers are of the same form as defined by the core SIP specification, which is defined as follows:

```
field-name: field-value *(;parameter-name[=parameter-value])
```

The "createParameterable" method takes a string in the above form as input and creates the appropriate "Parameterable" object.

createRequest—"SipFactory.createRequest" method is used to create a "SipServletRequest" interface instance; this is covered later in this chapter. A "SipServletRequest" object represents a SIP protocol request message that either has been received as an incoming request or, as in the case that this method is mostly used, is acting as a User Agent Client (UAC) application. To see how this method is used as part of a UAC or B2BUA application, the reader should take a look back at Chapter 3, which discusses these application roles. There are three variations on this method, each of which takes a different input parameter, which is relevant for differing usages. All three methods take in the SIP Application Session to be associated to the request (as defined by the "SipApplicationSession" interface) and a string that represents the type of SIP method to be created (e.g., INVITE). The variations occur in how the originator (SIP "From" header) and receiver (SIP "To" header) of the request are specified:

- The first variation takes two "Address" objects to represent the SIP "To" and "From" headers.

- The second variation takes two "URI" objects to represent the SIP "To" and "From" headers.

- The third variation takes two strings to represent the SIP "To" and "From" headers.

createSipURI—"SipFactory.createSipURI" method is used to create a "SipURI" interface instance, which will be covered later in this chapter. A "SipURI" instance represents a SIP URI as used extensively in the protocol that appears in numerous SIP headers and is generally of the form "sip:user_part@domain_part;parameters." This method takes two string parameters, which represent the "user" part (lefthand side of SIP URI—before "@") and the "host" part (righthand side of SIP URI—after "@") of a SIP URI.

createURI—"SipFactory.createURI" method is used to create a URI interface instance, which will be covered later in this chapter. It can be seen as similar to the previous example, with the exception that it is for a wider range of URI types as well as SIP, such as the "tel" URI scheme. This method takes a string as a parameter and determines the appropriate scheme by looking at the start of the input string.

Example:

```
@Resource SipFactory sipFactory;

SipApplicationSession sas = sipFactory.createApplicationSession();
SipServletRequest req =
sipFactory.createRequest(sas,"INVITE","sip:stoffe@sipservlet.net",
"sip:chris@sipservlet.net");
```

10.1.2 AuthInfo

The "AuthInfo" interface allows applications to set common authentication information in SIP requests that are generated when acting as a User Agent Client (UAC). The type of information included in a SIP request when challenged by a SIP 401 or 407 response code is described in detail in RFC 3261 [1]. The "AuthInfo" interface has a single method:

> *addAuthInfo*—"AuthInfo.addAuthInfo" method enables an application to configure appropriate authentication information to a SIP request when acting in the role of a User Agent Client (UAC). Once configured, the instance of the "AuthInfo" interface can be added to a SIP request using other API methods such as "SipServletRequest.addAuthHeader." The method has the following four parameters, which are used to specify the authentication credentials:

- The first parameter is of type "integer" and represents the status code of the challenge response, either a 401 or 407 SIP response code.

- The second parameter is a string representing the realm to which the authentication challenge/response belongs.

- The third parameter is a string specifying the user name to be used in the challenge response.

- The fourth parameter is a string specifying the password to be used in the challenge response.

Example:

```
AuthInfo info = sipFactory.createAuthInfo();
info.addAuthInfo(401, "sipservlet.net", "stoffe", "verysecret");
SipServletRequest req = resp.getSession().createRequest("PUBLISH");
req.addAuthHeader(401, info);
req.send();
```

10.1.3 SipSessionsUtil

The "SipSessionsUtil" interface was introduced in the latest version of the SIP Servlet architecture (Version 1.1). It provides a utility for converged applications (meaning both SIP Servlet/HTTP and SIP Servlet/JEE convergence, as discussed in Chapter 2) to obtain an existing SIP Application Session using a number of index mechanisms. The utility can be obtained by an application either from the Servlet Context attribute named "javax.servlet.sip.SipSessionsUtil" or can be injected using the Java @Resource annotation. Examples and details of both techniques were included in Chapter 2, which looked into usage of this utility. The "SipSessionsUtil" interface has the following methods that are available to converged applications:

> *getApplicationSessionById*—"SipSessionUtil.getApplicationSessionById" method is used by an application to retrieve a specific SIP Application Session object. A SIP Application Session inherently has a unique identifier

within an application instance. The identifier can be retrieved using the "SipApplicationSession.getId" method in the SIP Servlet API. This method takes a string as a parameter that represents the same unique identifier of a SIP Application Session as an index. For example, the application might have been given the unique identifier using third-party techniques, which resulted in it looking up the particular application instance using the unique identifier and the "SipSessionsUtil.getApplicationById" method.

getApplicationSessionByKey—"SipSessionUtil.getApplicationSessionByKey" method functionality is almost exactly the same as the previous method in that it retrieves a SIP Application Session instance for a converged application. The major difference is the key used for retrieval. This method also takes a string parameter, except this time it represents a key as generated by the @SipApplicationKey annotation. This method takes an additional Boolean value that provides semantics if the specified SIP Application Session does not exist. If set to the value "true," the SIP Application Session is automatically created; if not found and if set to "false," the SIP Application Session is not created.

getCorrespondingSipSession—"SipSessionUtil.getCorrespondingSipSession" convenience method is used to obtain a SIP protocol session that has specific semantics within a SIP Application Session. Two optionally supported SIP extensions are supported in the latest version of the SIP Servlet architecture (Version 1.1): Join [3] and Replaces [4]. Both contain SIP headers that correspond to semantic information that links to existing SIP protocol sessions. This method allows for the two SIP Sessions to be easily associated by looking at the appropriate SIP header and then finding the corresponding SIP Session. The method takes two parameters. The first is a SIP Session representing one of the SIP protocol sessions to be correlated. The second is the SIP header that the container should inspect when looking to select the corresponding SIP Session. If the application wishes to find a SIP Session based on a SIP "Join" header operation, the parameter has the value "Join." If the application wishes to find a SIP Session based on a SIP "Replaces" header operation, the parameter has the value "Replaces."

Example:

```
@Resource SipSessionsUtil util;

sipFactory.createApplicationSessionByKey("foo");
.
.
.
SipApplicationSession sas =
util.getApplicationSessionByKey("foo",true);
```

10.1.4　ConvergedHttpSession

The "ConvergedHttpSession" interface is also new to the latest version of the SIP Servlet architecture (Version 1.1). Its intention is to provide convenience functions to developers who are working purely in a SIP Servlet/HTTP converged container. A HTTP Servlet container has the concept of a HTTP Session, which represents a protocol interaction using the HTTP protocol in a way similar to how the SIP Session represents a SIP interaction in a SIP Servlet container. The interface actually extends "HttpSession" (javax.servlet.http.HttpSession) from the HTTP Servlet specification. Using this interface allows an application to gain access to "HttpSession" functionality from a converged container. As well as the methods that are automatically made available to the application through the casting of a "HttpSession" to a "ConvergedHttpSession" (see HTTP Servlet specification for more details), the following methods are defined:

> *encodeURL*—"ConvergedHttpSession.encodeURL" method is used by applications to create an encoded URL that is returned as a string value. The method has two variations that appear in the API, which take difference parameters for encoding the URL:
>
> - The first instance of the method takes a string as a parameter that represents the HTTP URL to be encoded. The container encodes the HTTP URL with the unique HTTP session identifier.
> - The second instance of the method takes a string object representing the relative path to the current Web-based application as well as a string representing the scheme (either "http" or "https"). This information is then used in the encoding process.
>
> *getApplicationSession*—"ConvergedHttpSession.getApplicationSession" method is used in the context of the HTTP Session to obtain the related SIP Application Session for the application instance within the converged container. If no SIP Application Session exists, then one is created and associated to the HTTP Session in the context of the converged container. Subsequent calls to this method would then return the existing SIP Application Session.

10.1.5　SipServletListener

In Chapter 3, a number of core concepts for application development and lifecycle were discussed in detail. One such concept was the mandatory inclusion of an "init" method in a compliant SIP Servlet application. The "init" method is invoked by the container on deployment of an application and can contain important logic that needs to be invoked before processing can take place. An application needs to know when this initialization period has been completed so that it can carry on functioning normally, for example, if it is one of multiple SIP Servlet classes in an application. SIP Servlet classes should implement the "SipServlet Listener" to learn when initialization has taken place. The listener interface has a single method:

> *servletInitialized*—The "SipServletListener.servletInitialized" method is invoked by a container when a SIP Servlet class initialize method has completed. The method passes a parameter of type "SipServletConextEvent." A "SipServletContextEvent" class returned by this listener method has a single method call:
>
> * *getSipServlet*—Returns the SIP Servlet that has just completed the initialization phase of application deployment.

Example:

```
Public void doGet(HttpServletRequest req, HttpServletResponse resp) {
 HttpSession hs = req.getSession();
if( hs.instanceof ConvergedHttpSession ) {
 //We are in a JSR 289 converged container
 ConvergedHttpSession chs = (ConvergedHttpSession) hs;
 SipApplicationSession sas = chs.getApplicationSession();
 Iterator<SipSession> ssList = getSessions("sip"); //Find the ongoing
conf
} else {
 //We cannot utilize SIP
}
```

10.2   Application Constructs

A SIP Servlet container, through its common API, provides clear boundaries and patterns for successful application development and interoperation. The constructs defined in this section are common across SIP Servlet-based applications and provide the foundation for the design pattern. Some of the main principles such as a SIP Application Session ("SipApplicationSession" interface) and a SIP Session ("SipSession" interface) have already been discussed in more detail in Chapter 3. They will be briefly covered again in this section to provide more depth in the context of the SIP Servlet API.

10.2.1   SipApplicationSession

The "SipApplicationSession" interface is a single representation of an application instance. It provides a container used for associating related SIP protocol sessions (see "SipSession" interface later in this section) and storing application-related data. You might recall that the book has discussed numerous mechanisms in varying deployments for creating new and retrieving existing SIP Application Session instances (e.g., using "SipFactory," "SipSessionsUtil," and "Converged HttpSession"). The "SipApplicationSession" interface has the following interface methods that can be used in an application:

encodeURL—"SipApplicationSession.encodeURL" method is used by applications to encode a specific URL with the unique application session identifier. The resulting URL can then be distributed using third-party mechanisms, and when it is received again by the container in the form of an HTTP request, it is able to decode the URL and associate the new HTTP Session ("HttpSession") with the existing SIP Application Session that was used to encode in the first place.

getApplicationName—"SipApplicationSession.getApplicationName" method returns the name of the SIP Application (".sar" file) to which this specific SIP Application Session belongs. As discussed in Chapter 3, the latest version of the SIP Servlet architecture (Version 1.1) requires applications specify a name for an application archive using either the "application-name" element in the deployment descriptor or the "name" element from the @Sip Application Java annotation.

getCreationTime—"SipApplicationSession.getCreationTime" method returns the time when the SIP Application Session was created.

getExpirationTime—"SipApplicationSession.getExpirationTime" method returns the time when the SIP Application Session is due to expire. As discussed Chapter 3, it is possible to set a SIP Application Session to be long lived and never expire. If this occurs, the "getExpirationTime" method returns 0.

getId—"SipApplicationSession.getId" method returns the unique indexing value that is created for every SIP Application Session. This unique identifier can be used in the future to reference a SIP Application Session instance, for example, using the "SipSessionsUtil" interface discussed in this section and described in Chapter 3.

getLastAccessedTime—"SipApplicationSession.getLastAccessedTime" method returns the time that the last SIP signaling interaction took place.

getSession—"SipApplicationSession.getSession" method returns a specific SIP protocol session as specified by the parameters of the request. This method has two parameters. The first specifies the unique identifier that is used to represent an instance of the SIP protocol session (an instance of the "SipSession" interface). As with instances of the "SipApplicationSession" interface, instances of the "SipSession" interface have a unique identifier value.

The second parameter indicates the type of protocol session that is to be retrieved. The possible values are "http" when in use with a converged SIP Servlet/HTTP container and "sip" for other types of container.

getSessions—"SipApplicationSession.getSessions" method returns a list of all protocol sessions associated with the SIP Application Session. They can be both SIP and HTTP depending on the string value passed in as the methods only parameter.

getSipSession—"SipApplicationSession.getSipSession" method returns an instance of a SIP protocol session ('SipSession interface instance) depending on the unique SIP Session identifier specified as a string in the only input parameter.

getTimer—"SipApplicationSession.getTimer" method returns a specific timer task. As with instances of both the "SipApplicationSession" and "Sip Session" interfaces, a timer task (represented by the "ServletTimer" interface detailed in this section) is represented by a unique identifier. The identifier, represented as a string value, is used as the only parameter to this method to obtain a specific timer task. Timer tasks are discussed in more detail in Chapter 2.

getTimers—"SipApplicationSession.getTimers" method returns all timer tasks (as represented by an instance of the "ServletTimer" interface) associated with the associated SIP Application Session.

invalidate—"SipApplicationSession.invalidate" method terminates the existence of a SIP Application Session. The instance of the "SipApplicationSession" interface can no longer be used or referenced by an application, and all associated application data is destroyed. Explicit invalidation was discussed in Chapter 3.

getInvalidateWhenReady—"SipApplicationSession.getInvalidateWhen-Ready" method is used by the container to indicate the "ready-to-invalidate" status. For a more detailed explanation of the "ready-to-invalidate" concept, take a look at Chapter 3. This method returns a Boolean value with the value of "true" indicates that the container is monitoring the "ready-to-invalidate" status of the specific "SipApplicationSession" interface instance, and a value of "false" indicates that the container is not monitoring the "ready-to-invalidate" status of the specific "SipApplicationSession" interface instance.

isReadyToInvalidate—"SipApplicationSession.isReadyToInvalidate" method returns a Boolean value indicating whether the "SipApplication Session" interface instance is in the "ready-to-invalidate" state. A value of "true" indicates that the "SipApplicationSession" interface instance is in the "ready-to-invalidate" state, and a value of "false" indicates that the "Sip

ApplicationSession" interface instance is not in the "ready-to-invalidate" state. For more information on the "ready-to-invalidate" state, see Chapter 3.

setInvalidateWhenReady—"SipApplicationSession.setInvalidateWhen Ready" method enables an application to set the "ready-to-invalidate" status of the container. The method has a single Boolean value parameter with a value of "true" indicating that container should monitor "ready-to-invalidate" status of a "SipApplicationSession" interface instance and a value of "false" indicating that the container should not monitor "ready-to-invalidate" status of a "SipApplicationSession" interface instance.

isValid—"SipApplicationSession.isValid" method returns a Boolean value indicating whether a SIP Application Session is still currently active. A returned value of "true" means that the SIP Application Session instance is still available, while a value of "false" means it has been invalidated.

getAttribute—"SipApplicationSession.getAttribute" method returns a Java object representation of application-level data that has previously been stored (e.g., using the "setAttribute" method covered later in this section). This book discussed storage of application data using the "SipApplicationSession" interface in Chapter 3. When storing application data, a unique name must be specified as an index for later referral. The "getAttribute" method has a single parameter of type string that indicates the unique index name of the application data to be retrieved and returned in Java object type.

setAttribute—"SipApplicationSession.setAttribute" method inserts an application data object in association with the "SipApplicationSession" interface instance. The unique indexing string is used in conjunction with the storage of the application data object so that it can be referenced in the future with related methods, such as "getAttribute" and "removeAttribute," which appear on the "SipApplicationSession" interface. The unique string identifier is passed as the only parameter in this method call.

getAttributeNames—"SipApplicationSession.getAttributeNames" method returns a list of strings indicating the names (as discussed in the "getAttribute" method) of the application data stored against a "SipApplicationSession" interface instance.

removeAttribute—"SipApplicationSession.removeAttribute" method removes an application data object that has previously been stored within a "SipApplicationSession" interface instance. As mentioned regarding the "getAttribute" method, application data is uniquely indexed using a string value. The unique identifier is passed as the only parameter to indicate which application data object should be removed from the "SipApplicationSession" interface instance.

setExpires—"SipApplicationSession.setExpires" method allows an application to extend the lifetime of a SIP Application Session. The method has a single parameter that takes an integer value representing the number of minutes to extend the lifetime. The method call also has a return value of the number of minutes that the SIP Application Session has been extended by, maybe due to container policy. An application can request that a SIP Application Session never expire by passing the value of "0" as the parameter. The topic of session expiry is discussed in more detail in Chapter 3.

Example:

```
SipApplicationSession sas = sipFactory.createApplicationSession();
URL myHttpServlet = sas.encodeURL(new URL( "http://sipservlet.net/
myServlet" ) );
// When a HttpSession is created using the URL it will be connected
to our SAS
```

10.2.2 SipApplicationSessionActivationListener

Application data objects stored in a SIP Application Session can use this listener interface to be notified by the container when the "SipApplicationSession" interface instance will be "passivated" or "activated." The interface has two methods:

sessionDidActivate—"SipApplicationSessionActivationListener.session DidActivate" method is invoked on the listener interface when an instance of the "SipApplicationSession" interface is activated, for example,after it has been migrated to another Java Virtual Machine or replicated for failover. The method provides a single class as a parameter, called "SipApplication SessionEvent." The "SipApplicationSessionEvent" class has a single method:

- *getApplicationSession*—"SipApplicationSessionEvent.getApplication Session" method returns the "SipApplicationSession" interface instance that has been activated.

sessionWillPassivate—"SipApplicationSessionActivationListener.session DidActivate" method is invoked on the listener interface when an instance of the "SipApplicationSession" interface is passivated, for example, after it has been migrated to another Java Virtual Machine or replicated for failover. The method provides a single class as a parameter called "SipApplication SessionEvent." The "SipApplicationSessionEvent" class has a single method:

- *getApplicationSession*—"SipApplicationSessionEvent.getApplication Session" method returns the "SipApplicationSession" interface instance that has been passivated.

10.2.3 SipApplicationSessionAttributeListener

This listener interface can be used to monitor when changes to application data objects are made that are associated with an instance of a "SipApplicationSession" interface. The listener interface has three methods that can be invoked on a change in application data:

attributeRemoved—"SipApplicationSessionAttributeListener.attribute Removed" method is invoked on the listener interface when an application data object is removed from a SIP Application Session (e.g., the "SipApplicationSession.removeAttribute" method could have been used). The invocation will pass in a parameter of type "SipApplicationSessionBindingEvent" class, which itself has a two methods:

- *getApplicationSession*—"SipApplicationSessionBindingEvent.get ApplicationSession" returns an instance of the "SipApplicationSession" interface from which the data object was removed.

- *getName*—"SipApplicationSessionBindingEvent.getName" returns the unique string index value used to identify application-level data.

attributeAdded—"SipApplicationSessionAttributeListener.attributeAdded" method is invoked on the listener interface when an application data object is added to a SIP Application Session (e.g., the "SipApplicationSession.set Attribute" method could have been used). The invocation will pass in a parameter of type "SipApplicationSessionBindingEvent" class, which itself has two methods:

- *getApplicationSession*—"SipApplicationSessionBindingEvent.get ApplicationSession" returns an instance of the "SipApplicationSession" interface to which the data object was added.

- *getName*—"SipApplicationSessionBindingEvent.getName" returns the unique string index value used to identify application-level data.

attributeReplaced—"SipApplicationSessionAttributeListener.attribute Replaced" method is invoked on the listener interface when an application data object is replaced from a SIP Application Session (e.g., the "Sip ApplicationSession.setAttribute" method could have been used). The invocation will pass in a parameter of type "SipApplicationSessionBindingEvent" class, which itself has a two methods:

- *getApplicationSession*—"SipApplicationSessionBindingEvent.get ApplicationSession" returns an instance of the "SipApplicationSession" interface of which the data object was replaced.

- *getName*—"SipApplicationSessionBindingEvent.getName" returns the unique string index value used to identify application-level data.

Example:

```
@SipListener

public class SniffServlet extends SipServlet implements
SipApplicationSessionAttributeListener {
 public void attributeAdded(SipApplicationSessionBindingEvent ev){
  log("Attribute added to SAS : "+ev.getName()+" =
"+ev.getApplicationSession().getAttribute( ev.getName() ) );
 }
}
```

10.2.4 SipApplicationSessionBindingListener

This listener interface is used to notify application data objects when they have been bound and unbound to an instance of the "SipApplicationSession" interface either programmatically or as a by-product of "SipApplicationSession" invalidation. The listener interface has the following two methods:

valueBound—"SipApplicationSessionBindingListener.valueBound" method is invoked when an application data object is bound to an instance of the "SipApplicationSession" interface (e.g., the "SipApplicationSession.set Attribute" method could have been used). The invocation will pass in a parameter of type "SipApplicationSessionBindingEvent" class, which itself has two methods:

- *getApplicationSession*—"SipApplicationSessionBindingEvent.get ApplicationSession" returns an instance of the "SipApplicationSession" interface to which the data object was bound.

- *getName*—"SipApplicationSessionBindingEvent.getName" returns the unique string index value used to identify application-level data.

valueUnbound—"SipApplicationSessionBindingListener.valueUnbound" method is invoked when an application data object is unbound from an instance of the "SipApplicationSession" interface (e.g., the "SipApplication Session.removeAttribute" method could have been used). The invocation will pass in a parameter of type "SipApplicationSessionBindingEvent" class, which itself has two methods:

- *getApplicationSession*—"SipApplicationSessionBindingEvent.get ApplicationSession" returns an instance of the "SipApplicationSession" interface from which the data object was unbound.

- *getName*—"SipApplicationSessionBindingEvent.getName" returns the unique string index value used to identify application-level data.

Example:

```
@SipListener

public class SniffServlet extends SipServlet implements
SipApplicationSessionBindingListener {
 public void valueUnbound(SipApplicationSessionBindingEvent event) {
  log("Attribute unbound : "+event.getName());
 }
}
```

10.2.5 SipApplicationSessionListener

The "SipApplicationSessionListener" interface provides the ability to receive notifications on the state of the underlying instance of the "SipApplicationSession" interface. The interface has the following methods that can be implemented:

>*sessionCreated*—"SipApplicationSessionListener.sessionCreated" method is invoked when an instance of the "SipApplicationSession" interface is created. The invocation will pass in a parameter of type "SipApplicationSession Event," which has a single method:
>
>   • *getApplicationSession*—"SipApplicationSessionEvent.getApplication Session" returns an instance of the "SipApplicationSession" interface that was created within the application.
>
>*sessionDestroyed*—"SipApplicationSessionListener.sessionDestroyed" method is invoked when an instance of the "SipApplicationSession" interface is destroyed. The invocation will pass in a parameter of type "SipApplication SessionEvent," which has a single method:
>
>   • *getApplicationSession*—"SipApplicationSessionEvent.getApplication Session" returns an instance of the "SipApplicationSession" interface that was destroyed within the application.
>
>*sessionExpired*—"SipApplicationSessionListener.sessionExpired" method is invoked when an instance of a the "SipApplicationSession" interface expires. The invocation will pass in a parameter of type "SipApplicationSession Event," which has a single method:
>
>   • *getApplicationSession*—"SipApplicationSessionEvent.getApplication Session" returns an instance of the "SipApplicationSession" interface that has expired within the application.
>
>*sessionReadyToInvalidate*—"SipApplicationSessionListener.sessionReady ToInvalidate" method is invoked when an instance of the "SipApplication Session" interface moves to the "ready-to-invalidate" state (as introduced

earlier in this section and discussed in more detail in Chapter 3). The invocation will pass in a parameter of type "SipApplicationSessionEvent," which has a single method:

- *getApplicationSession*—"SipApplicationSessionEvent.getApplication Session" returns an instance of the "SipApplicationSession" interface that transitioned to the "ready-to-invalidate" state within the application.

Example:

```
@SipListener

public class ExtServlet extends SipServlet implements
SipApplicationSessionListener {
 public void sessionExpired(SipApplicationSessionEvent ev) {
  ev.getApplicationSession().setExpires(5); //Extend 5 min
 }
}
```

10.2.6 SipSession

The "SipSession" interface is a representation of a SIP signaling interaction between users and is almost an identical state replication of a SIP dialog as defined in the core SIP specification [1]. In Chapter 3, we looked in more detail at the "Sip Session" interface construct and its association with the SIP protocol. This section will take a closer look at the interface and its associated methods:

createRequest—"SipSession.createRequest" method is used to create a new SIP protocol request as a SIP user agent within the context of a protocol interaction. For example, creating a subsequent re-INVITE or an UPDATE request in SIP within a SIP dialog would be achieved by calling this method. This method should not be confused with the sending of requests that have no prior signaling interaction and no SIP dialog (and therefore no instance of the "SipSession" interface). For such new interactions, such as initiating a call as a User Agent Client (UAC) application, the "SipFactory.createRequest" should be used. This method has a single string parameter that specifies the SIP method that is being created. The method returns an instance of the "SipServletRequest" interface, with the majority of SIP header fields auto-populated with legal mandatory values. The application can still manipulate the "SipServletRequest" object as required and then invoke the "SipServlet Request.send" method to dispatch the message. See the "SipServletRequest" section for more information on specific manipulation that can take place.

getApplicationSession—"SipSession.getApplicationSession" method call returns the associated SIP Application Session. The method takes no parameters and returns the appropriate instance of the "SipApplicationSession" interface.

getCallId—"SipSession.getCallId" method returns the SIP "Call-ID" header as defined in the core SIP specification [1]. The "Call-Id" header is used as part of the dialog identification process and is retuned in string form.

getLocalParty—"SipSession.getLocalParty" method call returns an "Address" interface instance taken from the SIP "From" header as defined in the core SIP specification [1] for locally generated requests. The SIP "From" header indicates the originator of the request.

getRemoteParty—"SipSession.getRemoteParty" method call returns an "Address" interface instance taken from the SIP "To" header as defined in the core SIP specification [1] for locally generated requests. The SIP "To" header indicates the destination of the request.

getCreationTime—"SipSession.getCreationTime" method returns the time when this particular instance of the "SipSession" interface was created.

getId—"SipSession.getId" method returns a string value representing the unique identifier for the instance of the "SipSession" interface. Earlier in this section, we introduced a unique identifier that was used to index instances of the "SipApplicationSession" interface. The "SipSession" interface has a similar concept whereby every instance of the "SipSession" interface has a unique identifier that can be used to explicitly reference at any time. This method call provides a string representation of the unique identifier.

getLastAccessedTime—"SipSession.getLastAccessedTime" method returns the last time that a client sent a request using this particular instance of the "SipSession" interface.

getRegion—"SipSession.getRegion" interface method is used to obtain the SIP routing region in which this instance of the "SipSession" interface is operating. The method returns an instance of the "SipApplicationRouting Region" class. The "SipApplicationroutingRegion" object will indicate a value of either "NEUTRAL_REGION," "ORIGINATING_REGION," or "TERMINATING_REGION." These values represent the region that the application is acting in for the SIP protocol interaction. For more information, take a look at Chapter 4.

getServletContext—"SipSession.getServletcontext" method returns the Servlet Context instance associated with the application. As discussed in the chapter "The SIP Servlet Container," the Servlet Context is associated with an application instance and is used to pass configuration values and offer utility instances such as SIP Factory and SIP Sessions utilities.

getState—"SipSession.getState" method returns the current SIP protocol state of the SIP dialog associated with the "SipSession" interface instance. A value of "INITIAL," "EARLY," "CONFIRMED," or "TERMINATED" is returned to the application. Apart from the INITIAL state, which represents a specific SIP Servlet architecture state and which is discussed in Chapter 3, the remainder are all mapped directly from the core SIP specification [1].

getSubscriberURI—"SipSession.getSubscriberURI" method returns an instance of the "URI" interface representing the subscriber that is being represented in the current invocation of an application. This is dependent on the region being serviced at the current time.

invalidate—"SipSession.invalidate" method is used to explicitly destroy the instance of the "SipSession" interface, including all application data objects stored. For more information on explicit invalidation, see Chapter 3.

isValid—"SipSession.isValid" method returns a Boolean value indicating whether the current instance of the "SipSession" interface has been invalidated; otherwise, the "SipSession" instance is considered valid. The value of "true" is returned when the instance of the "SipSession" interface is valid, and a value of "false" is returned when the instance of the "SipSession" interface has been invalidated.

getAttribute—"SipSession.getAttribute" method returns a Java object representation of application-level data that has previously been stored (e.g., using the "setAttribute" method to be covered later in this section). We discussed storage of application data using the "SipSession" interface in Chapter 3. When storing application data, a unique name must be specified as an index for later referral. The "getAttribute" method has a single parameter of type string that indicates the unique index name of the application data to be retrieved and returned in Java object type.

getAttributeNames—"SipSession.getAttributeNames" method returns a list of strings indicating the names (as discussed in the "getAttribute" method) of the application data stored against a "SipSession" interface instance.

removeAttribute—"SipSession.removeAttribute" method removes an application data object that has previously been stored within the "SipSession" interface instance. As mentioned in the "getAttribute" method, application data is uniquely indexed using a string value. The unique identifier is passed as the only parameter to indicate which application data object should be removed from the "SipSession" interface instance.

setAttribute—"SipSession.setAttribute" method inserts an application data object in association with the "SipSession" interface instance. The unique indexing string is again used in conjunction with the storage of the application data object so that it can be referenced in the future with related methods such as "getAttribute" and "removeAttribute," which appear on the "SipSession" interface. The unique string identifier is passed as the only parameter in this method call.

setHandler—"SipSession.setHandler" method allows an application to configure which SIP Servlet class should receive subsequent SIP protocol signaling for the "SipSession" interface instance (for the same SIP dialog). As we have discussed previously, an application (".sar") can consist of multiple SIP Servlets, and this enables an application to programmatically select which one should service an instance of the "SipSession" interface. The method has a single parameter of type string that specifies the name of the SIP Servlet class that will be invoked for incoming SIP messages for the "SipSession" interface instance.

getInvalidateWhenReady—"SipSession.getInvalidateWhenReady" method is used by the container to indicate the "ready-to-invalidate" status. For a more detailed explanation of the "ready-to-invalidate" concept, take a look at Chapter 3. This method returns a Boolean value whereby the value of "true" indicates that the container is monitoring the "ready-to-invalidate" status of the specific "SipApplicationSession" interface instance, and a value of "false" indicates that the container is not monitoring the "ready-to-invalidate" status of the specific "SipApplicationSession" interface instance.

isReadyToInvalidate—"SipSession.isReadyToInvalidate" method returns a Boolean value indicated if the "SipSession" interface instance is in the "ready-to-invalidate" state. A value of "true" indicates that the "SipSession" interface instance is in the "ready-to-invalidate" state, and a value of "false" indicates that the "SipSession" interface instance is not in the "ready-to-invalidate" state. For more information on the "ready-to-invalidate" state, see Chapter 3.

setInvalidateWhenReady—"SipSession.setInvalidateWhenReady" method enables an application to set the "ready-to-invalidate" status of the container. The method has a single Boolean parameter, with a value of "true" indicating the container should monitor "ready-to-invalidate" status of a "SipSession" interface instance and a value of "false" indicating that the container should not monitor "ready-to-invalidate" status of a "SipSession" interface instance.

setOutboundInterface—"SipSession.setOutboundInterface" method can be used in an environment in which multiple interfaces exist (multihomed). An application is able to select an interface for outgoing requests sent in association with a "SipSession" interface instance. The set of interfaces available to an application can be retrieved from the Servlet Context attribute "javax.servlet.sip.outboundInterfaces." On selecting an interface to use for outgoing requests, the container will automatically populate all the relevant

SIP headers with the appropriate, selected interface information (e.g., the SIP "Via" header). The method call accepts a single parameter of either type "java.net.InetSocketAddress" or "java.net.InetAddress."

Example:

```
SipSession ss = req.getSession();
if( ss.isReadyToInvalidate ) {
 ss.invalidate();
} else {
 ss.setInvalidateWhenReady(true);
 //The SS will be invalidated when SIP dialog is terminated
automatically.
}
```

10.2.7   SipSessionActivationListener

Application data objects stored in an instance of the "SipSession" interface can use this listener interface to be notified by the container when the "SipSession" instance will be "passivated" or "activated." The interface has two methods:

sessionDidActivate—"SipSessionActivationListener.sessionDidActivate" method is invoked on the listener interface when an instance of the "Sip Session" interface is activated, such as after it has been migrated to another Java Virtual Machine or replicated for failover. The method provides a single class as a parameter called "SipSessionEvent." The "SipSessionEvent" class has a single method:

- *getSession*—"SipSessionEvent.getSession" method returns the "Sip Session" interface instance that has been activated.

sessionWillPassivate—"SipSessionActivationListener.sessionDidActivate" method is invoked on the listener interface when an instance of the "Sip Session" interface is passivated, such as after it has been migrated to another Java Virtual Machine or replicated for failover. The method provides a single class as a parameter called "SipSessionEvent." The "SipSessionEvent" class has a single method:

- *getSession*—"SipSessionEvent.getSession" method returns the "SipSession" interface instance that has been passivated.

Example:

```
@SipListener

public class SniffServlet extends SipServlet implements
SipSessionActivationListener {
```

```
public void sessionWillPassivate(SipSessionEvent se) {
  //Clean up resources and store away non serializable properties
}
}
```

10.2.8   SipSessionAttributeListener

This listener interface can be used when changes to application data objects are made that are associated with an instance of a "SipSession" interface. The listener interface has three methods that can be invoked on a change in application data:

> *attributeRemoved*—"SipSessionAttributeListener.attributeRemoved" method is invoked on the listener interface when an application data object is removed from a "SipApplicationSession" interface instance (e.g., the "SipApplicationSession.removeAttribute" method could have been used). The invocation will pass in a parameter of type "SipSessionBindingEvent" class, which itself has a two methods:
>
> - *getSession*—"SipSessionBindingEvent.getSession" returns an instance of the "SipSession" interface in which the data object was removed.
> - *getName*—"SipSessionBindingEvent.getName" returns the unique string index value used to identify application-level data.
>
> *attributeAdded*—"SipAttributeListener.attributeAdded" method is invoked on the listener interface when an application data object is added to a "SipSession" interface instance (e.g., the "SipSession.setAttribute" method could have been used). The invocation will pass in a parameter of type "SipSessionBindingEvent" class, which itself has a two methods:
>
> - *getSession*—"SipSessionBindingEvent.getSession" returns an instance of the "SipSession" interface in which the data object was added.
> - *getName*—"SipSessionBindingEvent.getName" returns the unique string index value used to identify application-level data.
>
> *attributeReplaced*—"SipSessionAttributeListener.attributeReplaced" method is invoked on the listener interface when an application data object is replaced from a "SipApplicationSession" interface instance (e.g., the "SipSession.set Attribute" method could have been used). The invocation will pass in a parameter of type "SipSessionBindingEvent" class, which itself has a two methods:
>
> - *getApplicationSession*—"SipSessionBindingEvent.getSession" returns an instance of the "SipSession" interface in which the data object was replaced.

getName—"SipSessionBindingEvent.getName" method returns the unique string index value used to identify application-level data.

Example:

```
@SipListener

public class SniffServlet extends SipServlet implements
SipSessionAttributeListener {
 public void attributeReplaced(SipSessionBindingEvent event) {
  log("Attribute changed : "+event.getName()+" =
"+event.getSession().getAttribute( event.getName() ) );
 }
}
```

10.2.9   SipSessionBindingListener

This listener interface is used to notify application data objects when they have been bound and unbound to an instance of the "SipSession" interface either programmatically or as a by-product of "SipSession" invalidation. The listener interface has the following two methods:

> *valueBound*—"SipSessionBindingListener.valueBound" method is invoked when an application data object is bound to an instance of the "SipSession" interface (e.g., the "SipSession.setAttribute" method could have been used). The invocation will pass in a parameter of type "SipSessionBindingEvent" class which itself has a two methods:
> - *getApplicationSession*—"SipSessionBindingEvent.getSession" returns an instance of the "SipSession" interface in which the data object was bound.
> - *getName*—"SipSessionBindingEvent.getName" returns the unique string index value used to identify application-level data.
>
> *valueUnbound*—"SipSessionBindingListener.valueUnbound" method is invoked when an application data object is unbound from an instance of the "SipSession" interface (e.g., the "SipSession.removeAttribute" method could have been used). The invocation will pass in a parameter of type "SipSessionBindingEvent" class, which itself has a two methods:
> - *getApplicationSession*—"SipSessionBindingEvent.getSession" returns an instance of the "SipSession" interface in which the data object was unbound.
> - *getName*—"SipSessionBindingEvent.getName" returns the unique string index value used to identify application-level data.

Example:

```
@SipListener

public class SniffServlet extends SipServlet implements
SipSessionBindingListener {
 public void valueBound(SipSessionBindingEvent event) {
  log("New attribute bound : "+event.getName()+" =
"+event.getSession().getAttribute( event.getName() ) );
 }
}
```

10.2.10   SipSessionListener

The "SipSessionListener" interface provides the ability to receive notifications on the state of the underlying instance of the "SipSession" interface. The interface has the following methods that can be implemented:

> *sessionCreated*—"SipSessionListener.sessionCreated" method is invoked when an instance of the "SipSession" instance is created. The invocation will pass in a parameter of type "SipSessionEvent," which has a single method:
>
> - *getSession*—"SipSessionEvent.getSession" returns an instance of the "SipSession" interface that was created within the application.
>
> *sessionDestroyed*—"SipSessionListener.sessionDestroyed" method is invoked when an instance of the "SipSession" instance is destroyed. The invocation will pass in a parameter of type "SipSessionEvent," which has a single method:
>
> - *getApplicationSession*—"SipSessionEvent.getSession" method returns an instance of the "SipSession" interface that was destroyed within the application.
>
> *sessionReadyToInvalidate*—"SipSessionListener.sessionReadyToInvalidate" method is invoked when an instance of the "SipSession" instance moves to the "ready-to-invalidate" state (as introduced earlier in this section and discussed in more detail in Chapter 3). The invocation will pass in a parameter of type "SipSessionEvent," which has a single method:
>
> - *getApplicationSession*—"SipSessionEvent.getSession" returns an instance of the "SipSession" interface that transitioned to the "ready-to-invalidate" state within the application.

Example:

```
@SipListener

public class ConfServlet extends SipServlet implements
SipSessionListener {
```

```
public void sessionReadyToInvalidate(SipApplicationSessionEvent ev) {
  ev.getApplicationSession().setInvalidateWhenReady(false);
  //preventing the session to invalidate
}
}
```

10.3   SIP Message Routing

A large subset of SIP Servlet-based applications are used for routing SIP protocol messages in the role of a proxy server or as a B2BUA. The SIP Servlet API provides a number of abstracted interfaces that simplify such operations and automatically manage the underlying interactions with the associated SIP stack. The primary interfaces for such message routing are covered in the remainder of this section.

10.3.1   Proxy

A SIP proxy server is one of the most used constructs available in SIP networks, and operations associated with such a role are covered by the "Proxy" interface. The topic of using the "Proxy" interface was covered extensively Chapter 3. This section is intended to complement the explanation and examples provided to give more detail surrounding the associated method calls. The "Proxy" interface has the following methods:

> *cancel*—"Proxy.cancel" method call cancels the current proxy transaction and any child branches that have been created due to a forking operation. This results in a SIP CANCEL operation's being sent out on the appropriate branches, with the intention of a SIP final response being generated (SIP 4xx class response code). It should be noted that issuing this method call is only a request to cancel, and due to race conditions, an application should still expect to receive a success (SIP 2xx response code) final response. An overload version of the "Proxy.cancel" method also exists, which allows an application to indicate to the receiving endpoint the reason why it is canceling the request, as per RFC 3326 [5]. The overloaded "Proxy.cancel" method has three parameters. The first is a string parameter named "protocol," which is used to populate the source of the "cause" field in the SIP "Reason" header. The second parameter, named "reasonCode," is an integer value that is used to populate "cause" field in the SIP "Reason" header. Finally, a string parameter named "reasonText" is used to describe the reasoning for canceling the proxy operation. The following example demonstrates a SIP "Reason" header that would have been populated using this overloaded method call and associated parameters:

```
Reason: SIP ;cause=200 ;text="Call completed elsewhere"
```

The first parameter maps to the value "SIP," the second value maps to the value "200," and the third value maps to the value "Call completed elsewhere."

createProxyBranches—"Proxy.createProxyBranches" method creates a list (java.util.List) of the type "proxyBranch" interface from a provided set of SIP addresses. The "ProxyBranch" interface objects returned are then individually configurable, as opposed to a configuration change's impacting the whole proxy operation. The "ProxyBranch" interface is discussed later in this section. This method takes a list of "URI" interface objects, which is also covered in this chapter. The concept of being able to create and manipulate individual legs of a proxy operation was included in the latest version of the SIP Servlet architecture (Version 1.1). The topic is covered extensively in Chapter 3.

getProxyBranch—"Proxy.getProxyBranch" method enables an application to obtain an individual instance of the "ProxyBranch" interface. It can then apply individual branch-level configuration using the methods supplied on the "ProxyBranch" interface (which will be covered later in this chapter). The method has a parameter that is a "URI" interface instance that acts as a unique key for obtaining a "ProxyBranch" instance. The method returns the appropriate "ProxyBranch" interface instance to the requesting application.

getProxyBranches—"Proxy.getProxyBranches" method enables an application to obtain the list of "ProxyBranch" interface instances that are associated with a proxy operation (as in an instance of the "Proxy" interface). This method takes no parameters and returns a list of appropriate "ProxyBranch" interface objects.

startProxy—"Proxy.startProxy" method initiates the SIP protocol-level signaling associated with destinations previously added to the proxy list using the "Proxy.createBranches" method discussed in this section. As the "Proxy.createBranches" method can be called multiple times, the "Proxy.start Proxy" method can also be called to initiate any new SIP signaling branches that have been added to the set since the last time the method was called.

getAddToPath—"Proxy.getAddToPath" method indicates whether subsequent calls to the "Proxy.proxyTo" or "Proxy.startProxy" methods will result in a SIP "Path" header being inserted in the request as per RFC 3327 [6]. It should be noted that a SIP "Path" header will only be added to a request of type SIP REGISTER.

setAddToPath—"Proxy.setAddToPath" method indicates that, if a SIP REGISTER request is proxied, then it can add a SIP "Path" header, as specified in RFC 3327, to represent the container. The method has a single Boolean parameter, with a value of "true" indicating that a SIP "Path" header should be added to all SIP REGISTER proxy operations associated with this instance of the "Proxy" interface, and "false" indicating the "Path" header should not be added.

getPathURI—"Proxy.getPathURI" method provides the application with the "SipURI" interface instance that will be inserted on SIP REGISTER requests when configured using the "Proxy.setAddToPath" method. The method has no parameters and returns a "SipURI" interface object representing the SIP "Path" header from RFC 3327 [6].

getNoCancel—"Proxy.getNoCancel" method provides the container action that will be taken on receiving a SIP 2xx class response when multiple, alternative branches exist. The default SIP container behavior is to cancel the remaining branches. RFC 3841 allows for the default behavior to be overridden and remaining branches not to be canceled. This method returns a Boolean value, with "true" indicating the container will not cancel outstanding branches on receiving a SIP 2xx class response, and the value "false" indicating the container will cancel outstanding branches. This configuration is set using the "Proxy.setNoCancel" method.

setNoCancel—"Proxy.setNoCancel" method allows an application to override default SIP operation on receiving a SIP 2xx class response when multiple alternate branches exist. Default SIP container behavior on receiving a 2xx class response to a proxy operation is to automatically issue SIP CANCEL requests on the remaining branches. The "setNoCancel" method has a single Boolean parameter. If a value of "true" is specified, the container will not cancel remaining branches on receiving a SIP 2xx class response. If a value of "false" is specified, the container will cancel remaining branches on receiving a 2xx class response (the default is "false").

getOriginalRequest—"Proxy.getOriginalRequest" method allows the application to retrieve the original request that was received before it was manipulated and forwarded upstream. The method has no parameters and returns an instance of the "SipServletRequest" interface that represents a SIP request that was initially received from upstream of the container.

getParallel—"Proxy.getParallel" method provides the application with information on how the instance of the "Proxy" interface will handle multiple destinations—either forking in parallel (at the same time) or sequentially (one after another). The method will return a Boolean value, with "true" indicating that the instance of the "Proxy" interface is set to fork in parallel and "false" indicating that the instance of the "Proxy" interface is set to fork sequentially.

setParallel—"Proxy.setParallel" method enables an application to set whether multiple destinations should be forked either in parallel or sequentially. The method has a single Boolean parameter, with "true" indicating that the instance of the "Proxy" interface will proxy in parallel, while a value of "false" indicates that the instance of the "Proxy" interface will proxy sequentially.

setProxyTimeout—"Proxy.setProxyTimeout" method allows an application to specify a value for timeout of a proxy operation. The method has a single parameter that takes an integer value to indicate the number of seconds.

getProxyTimeout—"Proxy.getProxyTimeout" method returns an integer value representing the overall timeout of a proxy operation in seconds.

getRecordRoute—"Proxy.getRecordRoute" method indicates whether future SIP requests that are to be proxied will contain a SIP "Record-Route" header. The method returns a Boolean value, with "true" indicating that a SIP "Record-Route" header will be added to future proxied requests and "false" indicating that a SIP "Record-Route" header will not be added.

setRecordRoute—"Proxy.setRecordRoute" method instructs the container to add a SIP "Record-Route" header to an outgoing SIP request from an instance of the "Proxy" interface.

getRecordRouteURI—"Proxy.getRecordRouteURI" method returns the value that will be included in an outgoing SIP request in the SIP "Record-Route" header. The method returns an instance of the "SipURI" interface that represents the value that would be added to a request that is set to add a SIP "Record-Route" header using the "Proxy.setRecordRoute" method.

setRecurse—"Proxy.setRecurse" method specifies whether a container implementation will automatically recursively create new requests on receiving a SIP 3xx class response. The "setRecurse" method has a single parameter of Boolean value. A value of "true" indicates that the SIP Servlet container will recurse on receiving a 3xx class SIP response, while a value of "false indicates that the container will not recurse on receiving a 3xx class SIP response. In the case of "false," the 3xx class response is passed to the application by the container for further processing.

getRecurse—"Proxy.getRecurse" method indicates whether the instance of the "Proxy" interface is set to recurse on receiving a SIP 3xx class response. The method returns a Boolean value, with "true" indicating that "Proxy" interface instance is set to recurse on receiving a SIP 3xx class response and a value of "false" indicating that the "Proxy" interface instance is not set to recurse on receiving a SIP 3xx class response.

setSupervised—"Proxy.setSupervised" method specifies whether an instance of the "Proxy" interface is configured to receive SIP protocol responses related to the proxy operation. The method has a single parameter of Boolean value. A value of "true" indicates that SIP protocol responses should be passed to the application for further processing, and a value of "false" indicates that SIP protocol responses should not be passed to the application for further processing.

getSupervised—"Proxy.getSupervised" method indicates whether the instance of the "Proxy" interface is set to consume new SIP protocol responses. The method returns a Boolean value, with "true" indicating that "Proxy" interface instance is set to receive SIP protocol responses and a value of "false" indicating that the "Proxy" interface instance is not set to receive SIP protocol responses.

proxyTo—"Proxy.proxyTo" method is used to proxy a SIP request to the specified destination. The method has a single parameter value that indicates the location for the proxy request. The parameter either can be of interface type "URI" or can be a list of interface type "URI." The method can be invoked any number of times before a final SIP protocol response is passed upstream to add to the list of destinations.

setOutboundInterface—"Proxy.setOutboundInterface" method allows an instance of the "Proxy" interface to specify, from a list, a selected interface to use on multihomed machines. The use of this method will not only instruct the container to send from an interface but also impact the values a container places in key SIP headers such as "Record-Route" and "Via." The method call takes a single parameter that can be either of type "java.net.InetAddress" or can be of type "java.net.InetSocketAddress." A list of interfaces available to use by the application can be obtained from the Servlet Context attribute "javax.servlet.sip.outboundInterfaces."

Example:

```
Proxy p = req.getProxy();
p.setRecurse(true);
p.setRecordRoute(true);
p.proxyTo( req.getRequestURI() );
//Now the proxy is in for the duration of the dialog.
```

10.3.2 ProxyBranch

An instance of the "ProxyBranch" interface represents an individual branch in a proxy operation. It differs from the use of the "proxy" interface in that it allows more flexibility for applications specifically wanting to configure branches differently for a proxy operation. The following methods are defined on the "ProxyBranch" interface for individual branch configuration:

cancel—"ProxyBranch.cancel" method allows an application to cancel a specific branch of the proxy operation. Using "Proxy.cancel" will result in all branches being canceled, so this method provides a more specific level of branch control.

setAddToPath—"ProxyBranch.setAddToPath" method indicates that, if a SIP REGISTER request is proxied on the branch, then it will add a SIP "Path" header, as specified in RFC 3327, to represent the container. The method has a single Boolean parameter, with a value of "true" indicating that a SIP "Path" header should be added to the SIP REGISTER requests associated with this instance of the "ProxyBranch" interface and "false" indicating the "Path" header should not be added.

getAddToPath—"ProxyBranch.getAddToPath" method indicates whether a SIP "Path" header is being inserted in the request, as per RFC 3327 [6], on a specific "ProxyBranch" interface instance. It should be noted that a SIP "Path" header will be added only to a request of type SIP REGISTER.

getPathURI—"ProxyBranch.getPathURI" method provides the application with the "SipURI" interface instance that will be inserted on SIP REGISTER requests when configured using the "ProxyBranch.setAddToPath" method. The method has no parameters and returns a "SipURI" interface object representing the SIP "Path" header from RFC 3327 [6].

getProxy—"ProxyBranch.getProxy" method returns the "Proxy" interface instance associated with the "ProxyBranch" interface instance.

setProxyBranchTimeout—"ProxyBranch.setProxyBranchTimeout" method allows an application to specify a value for timeout of a specific proxy branch operation. The method has a single parameter that takes an integer value to indicate the number of seconds.

getProxyBranchTimeout—"ProxyBranch.getProxyBranchTimeout" method returns an integer value representing the timeout of a proxy branch operation in seconds.

setRecordRoute—"ProxyBranch.setRecordRoute" method instructs the container to add a SIP "Record-Route" header to an outgoing SIP request for an instance of the "ProxyBranch" interface.

getRecordRoute—The "ProxyBranch.getRecordRoute" method indicates whether future SIP requests that are to be proxied on the specific branch will contain a SIP "Record-Route" header. The method returns a Boolean value, with "true" indicating that a SIP "Record-Route" header will be added to proxied request and "false" indicating that a SIP "Record-Route" header will not be added.

getRecordRouteURI—"ProxyBranch.getRecordRouteURI" method returns the value that will be included in an outgoing SIP request in the SIP "Record-Route" header for the specific branch. The method returns an instance of the "SipURI" interface that represents the value that would be added to a request proxied on the specific branch.

setRecurse—"ProxyBranch.setRecurse" method specifies whether a container implementation will automatically recursively create new requests on receiving a SIP 3xx class response. The "setRecurse" method has a single parameter of Boolean value. A value of "true" indicates that the SIP Servlet container will recurse on receiving a 3xx class SIP response, while a value of "false" indicates that the container will not recurse on receiving a 3xx class SIP response. In the case of "false," the 3xx class response would be passed to the application by the container for further processing.

getRecurse—"ProxyBranch.getRecurse" method indicates whether the instance of the "ProxyBranch" interface is set to recurse on receiving a SIP 3xx class response. The method returns a Boolean value, with "true" indicating that "ProxyBranch" interface instance is set to recurse on receiving a SIP 3xx class response and a value of "false indicating that the "Proxy Branch" interface instance is not set to recurse on receiving a SIP 3xx class response.

getRecursedProxyBranches—"ProxyBranch.getRecursedProxyBranches" method returns a list of instances of the "ProxyBranch" interface that have resulted from a "ProxyBranch" interface instance's receiving a SIP 3xx class response.

getRequest—"ProxyBranch.getRequest" method returns the instance of the "SipServletRequest" interface associated with the "ProxyBranch" instance.

getResponse—"ProxyBranch.getResponse" method returns the last SIP response received on a branch, as represented by the instance of the "SipServletResponse" interface.

isStarted—"ProxyBranch.isStarted" method is used by an application to identify whether an instance of the "ProxyBranch" interface has been started (the "Proxy.startProxy" has been called). The method returns a Boolean value, with a value of "true" indicating that the instance of the "ProxyBranch" interface has been started and "false" indicating that the instance of the "Proxy Branch" interface has not been started.

setOutboundInterface—"ProxyBranch.setOutboundInterface" method allows an instance of the "ProxyBranch" interface to specify, from a list, a selected interface to use on multihomed machines. The use of this method will not only instruct the container to send from an interface but also impact the values a container places in key SIP headers such as "Record-Route" and

"Via." The method call takes a single parameter that can be of either type "java.net.InetAddress" or type "java.net.InetSocketAddress." A list of interfaces available for use by the application can be obtained from the Servlet Context attribute "javax.servlet.sip.outboundInterfaces."

Example:

```
List<SipURI> forks = new ArrayList<SipURI>();
forks.add( sipFactory.createSipURI("stoffe","sipservlet.net") );
forks.add( sipFactory.createSipURI("chris","sipservlet.net") );
List<ProxyBranch> branches = req.getProxy().createProxyBranches( forks );
branches.get(0).setRecordRoute(true); //Only Stoffe branch is Record-
Routed
req.getProxy().startProxy();
```

10.3.3 B2BuaHelper

Early versions of the SIP Servlet architecture soon identified many common types of roles that a typical application might assume. As discussed in Chapter 3, one of the most implemented was the Back-to-Back User Agent, or B2BUA. In the latest version of the SIP Servlet architecture (Version 1.1), a B2BUA helper class was introduced to abstract some of the common operations that occur in a B2BUA function. An instance of the B2BUA helper class can be obtained using the incoming SIP request to a container. The application would call the "SipServlet Request.getB2BuaHelper" method to gain an instance of the "B2BuaHelper" interface. Obtaining an instance of the "B2BuaHelper" interface signifies to the container that an application is going to act in the role of a B2BUA for this request (in the same way calling "SipServletRequest.getProxy" tells the SIP Servlet container that it will be acting as a proxy server). The following methods can then be used for B2BUA operations:

createCancel—"B2BuaHelper.createCancel" method allows an application to issue a SIP CANCEL request on outgoing legs of a B2BUA. The method has one parameter, which is the instance of the "SipSession" interface to be canceled.

createRequest—"B2BuaHelper.createRequest" method provides the ability for applications to generate and send requests acting as a B2BUA. There are three variations on the request, which all have differing roles to play.

• The first variation of this method is to be used for creating a new request as part of a B2BUA application. It takes in three parameters and returns an instance of the "SipServletRequest" interface. The first parameter specifies an instance of the "SipServletRequest" interface that is to be used for creating the outgoing leg. This involves maintaining appropriate SIP header values such as "To" and "From." The second parameter is a Boolean value that informs the container whether the two instances of the "SipSession" interface are linked. A value of "true" specifies that the new instance of the "SipSession" interface should be linked, and a value of "false" indicates it should not be linked. The term "linked" is a convenience that lets an application move between the two instances of the "SipSession" interface using the method call "B2BuaHelper.getLinkedSession," which is covered in this section. The third parameter is an optional "headerMap," which is a Java Map representing headers that should be copied from the original request to the new outgoing leg.

- The second variation of this method has only a single parameter, which specifies the instance of the "SipServletRequest" interface that is to be used for creating the outgoing leg.

- The third variation is used to create subsequent B2BUA requests based on the existing instance of the "SipSession" interface. It takes in three parameters and returns an instance of the "SipServletRequest" to be used for the subsequent B2BUA request. The first parameter is an instance of the "SipSession" interface to be used for sending the subsequent SIP request. The second parameter is an instance of the "SipServletRequest" interface representing the original request on which the new request should be based. The third parameter is an optional header map, which is a Java Map representing headers that should be copied from the original request to the new, subsequent request.

createResponseToOriginalRequest—"B2BuaHelper.createResponseToOriginal Request" method allows an application to generate multiple SIP response messages as a B2BUA to an incoming request, all of which have an individual instance of the "SipSession" interface. The method has three parameters and returns an instance of the "SipServletResponse" interface that is used to represent the new SIP response message that is to be sent. The first parameter is an instance of the "SipSession" interface that represents the original SIP request that is being responded to by the application. The second parameter is an integer representing the SIP status code that is to be included in the response message. The third parameter is a string value that allows an application to specify the reason phrase that can be included in the SIP response message.

getLinkedSession—"B2BuaHelper.getLinkedSession" method returns to an application the instance of the "SipSession" interface that was "linked" by the "B2BuaHelper" on using the "B2BuaHelper.createRequest" or the explicit "B2BuaHelper.linkSipSession" method. The method has a single parameter, which is the "SipSession" interface instance that is the known half of the two linked sessions.

getLinkedSipServletRequest—"B2BuaHelper.getLinkedSipServletRequest" method returns to an application the instance of the "SipServletRequest" interface that was "linked" by the "B2BuaHelper" on using the "B2Bua Helper.createRequest" method. The method has a single parameter, which is the "SipServletRequest" interface instance that is the known half of the two linked requests.

getPendingMessages—"B2BuaHelper.getPendingMessages" method returns a list of "SipServletMessage" instances that are considered uncommitted. The method takes two parameters. The first is the instance of the "SipSession" interface in question. The second is an ENUM value of either "UAC" or "UAS," signifying the type of role that the query is associated with.

linkSipSession—"B2BuaHelper.linkSipSession" method is the explicit mechanism that enables an application to link two instances of the "SipSession" interface for B2BUA functionality (it can also be achieved implicitly using the "B2BuaHelper.createRequest" method). The method takes two parameters, which represent the two "SipSession" instances that are being linked.

unlinkSipSession—"B2BuaHelper.unlinkSipSession" method is the explicit mechanism that enables an application to unlink two instances of the "Sip Session" interface that have been linked for B2BUA functionality. The method takes a single parameter that represents the "SipSession" instance that is to be unlinked.

Example:

```
SipServletRequest legA = req;
B2buaHelper b2b = req.getB2buaHelper();
SipServletRequest legB = b2b.createRequest(legA);
legB.setContent( legA.getContent(), legA.getContentType() );
legB.send();
```

10.3.4 SipErrorListener

The "SipErrorListener" is used to inform applications when specific SIP transactional behavior has not been completed properly. The listener interface has the following two methods that are invoked appropriately:

noAckReceived—"SipErrorListener.noAckReceived" listener method is invoked when an application in the role of a UAS does not receive a SIP ACK method after it has forwarded a final response to a SIP INVITE upstream. The method returns a parameter of type "SipErrorEvent." The "SipErrorEvent" class returned has two methods:

- *getRequest*—"SipErrorEvent.getRequest" method returns the "Sip ServletRequest" (the original request that generated the SIP response) interface instance associated with the error event.

- *getResponse*—"SipErrorEvent.getResponse" method returns the "SipServletResponse" (the SIP response) interface instance associated with the error event.

noPrackReceived—"SipErrorListener.noPrackRecevied" listener method is invoked when an application in the role of a UAS does not receive a SIP PRACK method after it has forwarded a reliable provisional response upstream. The method returns a parameter of type "SipErrorEvent." The "SipErrorEvent" class returned has two methods:

- *getRequest*—"SipErrorEvent.getRequest" method returns the "Sip ServletRequest" (the original request that generated the SIP reliable provisional response) interface instance associated with the error event.

- *getResponse*—"SipErrorEvent.getResponse" method returns the "Sip ServletResponse" (the SIP response) interface instance associated with the error event.

Example:

```
@SipListener

public class ErrorSipServlet extends SipServlet implements
SipErrorListener{
 public void noAckReceived(SipErrorEvent ee) {
  SipSession ss = ee.getResponse().getSession();
  SipServletRequest bye = ss.createRequest("BYE");
  bye.send();
 }
}
```

10.4   SIP Messaging Constructs

Underneath the entire higher level role constructs and related routing operations that were discussed in the previous section lie the actual SIP-level representations. These are the objects that are used to carry the SIP semantic detail, such as requests, responses, and SIP URIs. The remainder of this section will take a closer look at the interfaces that allow an application to gain access to the low-level SIP protocol machinery.

10.4.1   SipServletMessage

The "SipServletMessage" interface represents a SIP protocol message. The interface defines a number of common operations that are common across both SIP

protocol requests (represented by the "SipServletRequest" interface) and SIP protocol responses (represented by the "SipServletResponse" interface). This allows
a common interface for SIP Servlet applications to operate on both types of SIP
interactions at the message level. The following methods are defined on the
"SipServletMessage" interface for common SIP message manipulation:

addAcceptLanguage—"SipServletMessage.addAcceptLanguage" method will
insert a SIP "Accept-Language" header to the message. It takes a single
parameter of type "java.util.locale."

addAddressHeader—"SipServletMessage.addAddressHeader" method enables
an application to add a SIP header that complies with the "Address" interface format (as specified in this section). The method has three parameters.
The first is a string representation of the SIP header that is to be inserted.
The second is the "Address" interface instance, which will make up the
value of the SIP header. The third parameter is a Boolean value that specifies whether the "Address" interface instance is added as the first or last
value of the specified SIP header field. A value of "true" specifies that the
"Address" interface instance is added as the first value of the specified SIP
header, while a value of "false" specifies that it will be last.

addHeader—"SipServletMessage.addHeader" method adds a SIP header
and associated value to the SIP message. The method has two parameters.
The first parameter is a string value representing the name of the SIP header
to be added. The second parameter is a string value representing the value
of the SIP header.

addParameterableHeader—"SipServletMessage.addParameterable" method
adds a SIP header to a message of the format specified by the "Parameterable"
interface (which is covered in this section). The method has three parameters.
The first parameter is a string representing the name of the SIP header to be
added. The second parameter is the value to be inserted into the SIP header
and is of interface type "Parameterable" (which is covered in this section). The
third parameter is a Boolean value that specifies whether the "Parameterable"
interface instance is added as the first or last value of the specified SIP header
field. A value of "true" specifies that the "Parameterable" interface instance is
added as the first value of the specified SIP header, while a value of "false" specifies that it will be last.

getAcceptLanguage—"SipServletMessage.getAcceptLanguage" method returns
the locale of the SIP user agent that generated the "SipServletMessage" interface instance, as represented by the SIP "Accept-Language" header. The
locale is returned in the form "java.util.Locale."

getAcceptLanguages—"SipServletMessage.getAcceptLanguages" method
returns a list of the SIP user agents locales that generated the "SipServlet

Message" interface instance, as represented by the SIP "Accept-Language" header. The list of locales is returned in the form "java.util.Locale."

getAddressHeader—"SipServletMessage.getAddressHeader" method returns an instance of the "Address" interface associated with a SIP header. The method has a single string parameter that specifies the name of the SIP header whose value is being extracted.

getAddressHeaders—"SipServletMessage.getAddressHeaders" method returns a list containing instances of the "Address" interface associated with a SIP header. The method has a single string parameter that specifies the name of the SIP header whose value is being extracted.

getApplicationSession—"SipServletMessage.getApplicationSession" method returns the instance of the "SipApplicationSession" interface associated with this particular SIP message. The method has an optional variation that takes a Boolean parameter indicating whether a new instance of the "Sip ApplicationSession" interface should be created if it doesn't already exist. A value of "true" specifies that the container should create a new instance of the "SipApplicationSession" interface if it doesn't already exist, while a value of "false" specifies that it shouldn't be created.

setAttribute—"SipServletMessage.setAttribute" method, as with both the "SipApplicationSession" and "SipSession" interfaces, allows application data to be stored as attributes. Each piece of application data is stored as a Java object and is referenced using a unique index key of type string. This method has two parameters. The first parameter is the unique string index name for referencing the application data. The second parameter is the Java object representation of the application data. It should be noted that reusing the same index key for storing application data objects results in an over-write operation.

getAttribute—"SipServletMessage.getAttribute" method allows an application to retrieve an application data object that has been stored in the "SipServlet Message" interface. A single parameter of type string representing the unique index name of the application data object is used to obtain the Java object representation of the application data.

getAttributeNames—"SipServletMessage.getAttributeNames" method allows an application to retrieve a list of the unique application data keys associated with the associated "SipServletMessage" interface instance.

removeAttribute—"SipServletMessage.removeAttribute" method enables an application to delete data stored in the "SipServletMessage" interface instance. The method has a single parameter of type string that specifies the unique index name of the application data that is to be removed.

getCallId—"SipServletMessage.getCallId" method returns the SIP "Call-ID" header from the SIP message. The value is returned in a string representation.

getCharacterEncoding—"SipServletMessage.getCharacterEncoding" method returns the name of the charset (system for encoding a sequence of characters) used for the MIME body of the SIP message. The encoding type is returned in a string representation.

getContent—"SipServletMessage.getContent" method returns an object representation of the payload of the "SipServletMessage" interface instance.

getRawContent—"SipServletMessage.getRawContent" method returns a byte array representation of the payload of the "SipServletMessage" interface instance.

getContentLanguage—"SipServletMessage.getContentLanguage" returns the locale of the SIP message as represented by the SIP "Content-Language" SIP header (or a value of null if the SIP "Content-Langauge" header is not present). The locale is returned as type "java.util.Locale."

getContentLength—"SipServletMessage.getContentLength" method returns an integer representing the number of bytes making up the SIP message payload. This value is represented in the SIP "Content-Length" SIP header.

getContentType—"SipServletMessage.getContentType" method returns a string representation of the payload type of the "SipServletMessage" interface instance. The value is taken from the SIP "Content-Type" header.

getExpires—"SipServletMessage.getExpires" header method returns an integer value representing the SIP "Expires" header. A value of "–1" is returned if the header does not exist in the SIP Message.

getFrom—"SipServletMessage.getFrom" method returns the value of the SIP "From" header. The value is returned as an instance of the SIP "Address" interface.

getHeader—"SipServletMessage.getHeader" method returns a string representation of the value of a SIP header. The method has a single string parameter, which specifies the SIP header to be returned by the method.

getHeaderNames—"SipServletMessage.getHeaderNames" method returns to the application a list containing all of the SIP headers that appear in the "SipServletMessage" interface instance.

getHeaders—"SipServletMessage.getHeaders" method returns a list of SIP header values in string format. Some SIP headers are allowed to appear multiple times within a SIP message or can have comma-separated values. Using this method allows an application to retrieve the multiple values of a SIP header in a single call. The method has a single string parameter, which is a string representation of the SIP header name.

getInitialRemoteAddr—"SipServletMessage.getInitialRemoteAddr" method returns a string representation of the IP address relating to the upstream or downstream physical entity that routed the SIP message to the container regardless of container application routing.

getInitialRemotePort—"SipServletMessage.getInitialRemotePort" method returns a string representation of the port relating to the upstream or downstream physical entity that routed the SIP message to the container regardless of container application routing.

getInitialTransport—"SipServletMessage.getInitialRemoteTransport" method returns a string representation of the transport type relating to the upstream or downstream physical entity that routed the SIP message to the container regardless of container application routing.

getRemoteAddr—"SipServletMessage.getRemoteAddr" method returns a string representation of the IP address from which the SIP message was received. If the message was routed internally due to application composition, then the container will return the local IP address of the containers interface. To always retrieve the original physical remote address, the application should use the "SipServletMessage.getInitialRemoteAddr" method.

getRemotePort—"SipServletMessage.getRemotePort" method returns an integer representation of the port from which the SIP message was received. If the message was routed internally due to application composition, then the container will return a valid local port on the "Containers" interface. To always retrieve the original physical remote port, the application should use the "SipServletMessage.getInitialRemotePort method."

getRemoteUser—"SipServletMessage.getRemoteUser" method returns a string representation of the user attempting to send the SIP message represented by the "SipServletMessage" instance. If the user has not been authenticated, then null is returned.

getTransport—"SipServletMessage.getTransport" method returns a string representation of the protocol on which the SIP message was received. If the message was routed internally due to application composition, then the container will return null. To always retrieve the original physical transport, the application should use the "SipServletMessage.getInitialTransport method."

getLocalAddr—"SipServletMessage.getLocalAddr" method returns a string representation of the IP address relating to the upstream or downstream signaling entity that routed the SIP message. If the message was routed due to application composition, the value returned will equal an internal representation.

getLocalPort—"SipServletMessage.getLocalPort" method returns a string representation of the port relating to the upstream or downstream signaling entity that routed the SIP message. If the message was routed due to application composition, the value returned will equal an internal representation.

getMethod—"SipServletMessage.getMethod" method returns a string representation of the SIP protocol messages type, such as "INVITE."

setParameterableHeader—"SipServletMessage.setParameterableHeader" method allows an application to specify a SIP header of the form defined by the "Parameterable" interface. The method takes two parameters. The first parameter is a string representation of the SIP header name. The second parameter is an instance of the "Parameterable" interface (which is also defined in this section).

getParameterableHeader—"SipServletMessage.getParameterableHeader" method returns the instance of the "Parameterable" interface associated with a SIP header. The method takes a single string parameter, which specifies the SIP header to be returned in a "Parameterable" interface instance.

getParameterableHeaders—"SipServletMessage.getParameterableHeaders" method returns a list of instances of the "Parameterable" interface for a given SIP header. The method has a single parameter, which is a string representation of the header that is to be returned in a list of instances of the "Parameterable" interface.

getProtocol—"SipServletMessage.getProtocol" method returns the name and version of the protocol used in the "SipServletMessage" interface instance. This method will always return a string of the form "SIP/2.0."

getSession—"SipServletMessage.getSession" method returns the instance of the "SipSession" interface associated with the SIP protocol message interaction. The method has an optional Boolean parameter, which indicates whether an instance of the "SipSession" interface should be created if it doesn't already exist. A value of "true" specifies that the instance of the "SipSession" should be created if it doesn't exist and a value of "false" if it shouldn't be created.

getTo—"SipServletMessage.getTo" method returns an instance of the "Address" interface representing the SIP "To" header from the SIP protocol message.

getUserPrincipal—"SipServletMessage.getUserPrincipal" method returns an object of type "java.security.Principal" indicating the principal user. Take a look at the appropriate Java documentation for more information of the role of this object in the Java security framework.

isCommitted—"SipServletMessage.isCommitted" method returns a Boolean value indicating whether the instance of the "SipServletMessage" interface is committed. A value of "true" indicates the instance is committed, while a value of "false" indicates that it is not. If an instance of the "SipServlet Message" interface is deemed to be in the committed state, then it can no longer be modified. SIP Servlet 1.1 [7] specifies the states that indicate that an instance of the "SipServletMessage" interface should transition to a committed state:

- This message is an incoming request for which a final response has already been generated.
- This message is an outgoing request that has already been sent.
- This message is an incoming nonreliable provisional response received by a Servlet acting as a UAC.
- This message is an incoming reliable provisional response for which PRACK has already been generated. (Note that this scenario applies to containers that support the 100rel extension.)
- This message is an incoming final response received by a Servlet acting as a UAC for a nonINVITE transaction.
- This message is a response that has been forwarded upstream.
- This message is an incoming final response to an INVITE transaction, and an ACK has been generated.
- This message is an outgoing request, the client transaction has timed out, no response was received from the UAS, and the container generates a 408 response locally.

isSecure—"SipServletMessage.isSecure" method returns a Boolean value indicating whether the SIP protocol message received in creating the instance of the "SipServletMessage" interface was received over a secure transport protocol, such as TLS. A value of "true" indicates that the SIP protocol message was received over a reliable transport protocol, and a value of "false" conveys that the SIP protocol message was not.

isUserInRole—"SipServletMessage.isUserInRole" method returns a Boolean value indicating whether an authenticated user is authorized to assume a specific role within the application as defined in the deployment descriptor. The method has a single parameter of type string that indicates the role name being checked. A value of "true" indicates that the authenticated user does belong to a given role while a value of "false" indicates the user does not belong.

removeHeader—"SipServletMessage.removeHeader" method removes a SIP header from the SIP protocol message that the instance of the "SipServlet Message" interface represents. The method has a single parameter of type string that represents the name of the SIP protocol header to be removed.

send—"SipServletMessage.send" method is used to initiate the SIP protocol sending (dispatch of protocol packet to the SIPStack and network) of the

"SipServletMessage" instance when acting as either a User Agent Client (UAC) or a User Agent Server (UAS).

setAcceptLanguage—"SipServletMessage.setAcceptLanguage" method defines the preferred locale that this user is operating in. This helps with language-specific constructs such as reason phrases in SIP protocol responses and SIP "Warning" headers. The result of setting this method is that the SIP "Accept-Language" header is set to the appropriate language. The method has a single parameter of type "java.util.locale," which is explained in more detail in appropriate Java language documentation.

setAddressHeader—"SipServletMessage.setAddressHeader" method sets the appropriate SIP header to a value represented by the "Address" interface defined later in this chapter and discussed earlier in the book. The method has two parameters. The first parameter is of type string and represents the SIP header that is to be set. The second value is an instance of the "Address" interface, representing the value to be inserted into the previously specified SIP header.

setCharacterEncoding—"SipServletMessage.setCharacterEncoding" method sets the encoding to be used when converting the body of a "SipServlet Message" interface instance. The method has a single value of type string that specifies the character encoding to be used. Setting this method has impact when attempting to manipulate the body of a "SipServletMessage" interface instance using methods such as "SipServletMessage.setContent" and "SipServletMessage.getContent" (which are also described in this section).

setContent—"SipServletMessage.setContent" method is used to set the body of a SIP protocol message, which is represented by an instance of the "SipServletMessage" interface. The method has two parameters. The first is of type "java.lang.Object," which is simply a Java object representing the SIP message body. The second parameter is of string and specifies the MIME type of the SIP message body.

setContentLanguage—"SipServletMessage.setContentLanguage" method sets the locale of a SIP protocol message that is represented by an instance of the "SipServletMessage" interface. This method setting impacts SIP headers such as "Content-Language" and "Content-Type." It takes a single parameter of type "java.util.Locale," which is defined in the relevant Java documentation.

setContentLength—"SipServletMessage.setContentLength" method sets a value to the SIP "Content-Length" header as represented by the instance of the "SipServletMessage" interface. Manually setting this header is not recommended, as using the "SipServletMessage.setContent" method ensures that the correct value is set to represent the SIP message body. The method has a single parameter of type integer that specifies the value to be used in the SIP "Content-Length" header.

setContentType—"SipServletMessage.setContentType" method sets the content type of a SIP protocol message represented by the "SipServletMessage" interface. The method has a single parameter of type string that represents the MIME type of the content contained in the SIP protocol message.

setExpires—"SipServletMessage.setExpires" method is used to set the value in an instance of the "SipServletMessage" interface of the "Expires" header. The method has a single parameter of type integer that represents the number of seconds to be inserted into the SIP "Expires" header.

setHeader—"SipServletMessage.setHeader" method allows an application to set a SIP protocol header that is represented in an instance of the "SipServletMessage" interface. The method has two parameters. The first is of type string and represents the name of the SIP header to be set. The second parameter is also of type string and sets the value for the SIP header represented by the first parameter in this method call.

getHeaderForm—"SipServletMessage.getHeaderForm" method returns the form that the SIP headers in a message would be rendered once out on the network. SIP 2.0 [1] allows some headers to have a short format. An example of one such header is the "From" SIP header, which can have the short form representation "f."

setHeaderForm—"SipServletMessage.setHeaderForm" method can be used to specify whether the long header format or the short header format should be used. (according to SIP 2.0 [1]). If the short format is set, then all headers that have a short format defined in an RFC would use that, while unknown headers or headers without a short format specified would use the long format.

Example:

```
//Nice helper function
public void copyBody( SipServletMessage source, SipServletMessage dest )
{
 dest.setContent( source.getContent(), source.getContentType );
 dest.setContentLanguage( source.getContentLanguage() );
 dest.setCharacterEncoding( source.getCharacterEncoding() );
}
```

10.4.2   SipServletRequest

The "SipServletRequest" interface represents a SIP request message (as opposed to a SIP response message). The "SipServletRequest" interface extends the "javax.servlet" interface to provide SIP specific methods and also implements the "SipServletMessage" interface to provide common methods across both "Sip

ServletRequest" and "SipServletResponse" interfaces. The following methods are defined for specific manipulations on a "SipServletRequest" interface instance:

addAuthHeader—"SipServletRequest.addAuthHeader" provides a convenience method for applications acting as a User Agent Client that are challenged for authentication using a SIP 401 or 407 response code (see RFC 3261 [1] for more detail on the call flow and headers). There are two variations of the method:

- The first variation of the "SipServletRequest.addAuthHeader" has two parameters. The first parameter is the instance of the "SipServlet Response" interface instance that triggered the action. The second parameter is an instance of the "AuthInfo" interface that acts as a container for appropriate authentication credentials (e.g., user name and password). Detailed information describing the "AuthInfo" interface is included in this chapter.

- The second variation of the "SipServletRequest.addAuthHeader" has three parameters that enable an application acting as a UAC to create appropriate authentication headers in a request without using the "AuthInfo" interface instance. The first parameter is the instance of the "SipServletResponse" interface instance that triggered the action. The second parameter is of type string and represents the user name for the authentication operation. The third parameter is also of type string and represents the password for the authentication operation.

createCancel—"SipServletRequest.createCancel" method returns an instance of the "SipServletRequest" interface for canceling a SIP INVITE request when acting as a User Agent Client. An application acting as a User Agent Client can then invoke the "SipServletRequest.send" method on the returned instance to cancel.

createResponse—"SipServletRequest.createResponse" method can be used by an application on receiving a SIP request. Calling this method allows the application to generate a SIP response, which can then be sent using the "SipServletRequest.send" method. Generating and sending such a response is an indication to the container that the application is acting in the User Agent Server role. There are two variations of this method:

- The first variation has only a single parameter of type integer. The integer value passed in represents the status code of the SIP response message (e.g., "200" would create a SIP "200 OK" response).

- The second variation has two parameters. The first parameter is equal to the parameter introduced in the first variation of this method and is of type integer to represent the SIP status code response that is to be generated. The second parameter is of type string and specifies the textual reason phrase, which is included in first line of the SIP response along with the status code.

getB2buaHelper—"SipServletRequest.getB2buaHelper" method returns an instance of the "B2buaHelper" interface. Using this method indicates to the SIP Servlet container that the application wishes to act in the role of a B2BUA. Details of the functionality provided by the "B2buaHelper" interface are included in this section.

getInitialPoppedRoute—"SipServletRequest.getInitialPoppedRoute" method returns the original SIP "Route" header that was popped before being presented to applications. Sometimes the contents of the SIP "Route" header are used to provide contextual information, so it's important that the contents be made available to applications [e.g., some implementations of the IP Multimedia Subsystem (IMS) use the SIP "Route" header to carry appropriate information]. The method returns an instance of the "Address" interface, which is described in detail in this section. If no SIP "Route" header has been popped for the request, then null is returned.

getMaxForwards—"SipServletRequest.getMaxForwards" method returns an integer value representing the SIP protocol "Max-forwards" header. This value is used by forwarding entities to guard against messages looping. Each forwarding entity should decrement the value of this header by one, and if it reaches zero, the appropriate SIP error response (483) will be generated instead of forwarding.

getPoppedRoute—"SipServletRequest.getPoppedRoute" method returns the top header removed by the container before being dispatched to a SIP servlet application. This differs from the "SipServletRequest.getInitialPoppedRoute" method, which preserves the SIP "Route" header that was present when the SIP request entered the container from the network. The instance of the "Address" interface that is retuned by this method represents the SIP "Route" header popped as a result of potentially multiple application invocations with popped headers (e.g., the Application Router can push headers that result in invocation).

getProxy—"SipServletRequest.getProxy" method returns an instance of the "Proxy" interface that is used when an application wishes to act in the role of a SIP proxy server. See the information provided in this section on the "Proxy" interface for various proxy server configurations that are provided.

getRegion—"SipServletRequest.getRegion" method provides the region in which a SIP request has been invoked. An Application Router will generally dispatch requests to applications while acting in a specific region. This method returns an instance of the "SipApplicationRoutingRegion" class. The "SipApplicationRoutingRegion" object will indicate a value of either "NEUTRAL_REGION," "ORIGINATING_REGION," or

"TERMINATING_REGION." These values represent the region that the application is acting in for the SIP protocol interaction. For more information, take a look at Chapter 4.

getRequestURI—"SipServletRequest.getRequestURI" method returns an instance of the "URI" interface representing the SIP Request URI that appears in the first line of a SIP request. The "URI" interface instance represents the current destination of the SIP request.

getRoutingDirective—"SipServletRequest.getRoutingDirective" method returns the ENUM "SipApplicationRoutingDirective," which indicates the routing directive of a request. A request is processed and sequenced based on the routing directive, which has the following values:

- *NEW*—Indicates that the request is new and should be treated as if it has never been seen before, from an application sequencing perspective.

- *CONTINUE*—Indicates that the request is not new and is being processed in an existing application sequence.

- *REVERSE*—Indicates that the request is being processed in the reverse direction of an existing application sequence.

An application can obtain the value of the Routing Directive from the "SipApplicationRoutingDirective" ENUM by invoking the "SipApplication RoutingDirective.valueOf" method, which returns the appropriate string representation.

setRoutingDirective—"SipServletRequest.setRoutingDirective" method enables an application to set the previously introduced Routing Directive for an outgoing request. The method takes two parameters. The first parameter is an instance of the "SipApplicationRoutingDirective," which was introduced in the definition of the "SipServletRequest.getRoutingDirective" method. It will contain a value of either "NEW," "CONTINUE," or "REVERSE." The second parameter is the instance of the "SipServlet Request" interface that the Routing Directive is being set for. Setting the Routing Directive impacts sequencing behavior of the outgoing request. The Routing Directive has default values that depend on the role in which the application is acting. An application uses this method to override default behavior.

getSubscriberURI—"SipServletRequest.getSubscriberURI" method returns an instance of the "URI" interface representing the user that is currently being serviced by the container.

isInitial—"SipServletRequest.isInitial" method informs applications whether the incoming request is an initial or a subsequent SIP request. This method is of type Boolean, with a value of "true" indicating that a request is initial and a value of "false" indicating that it is a subsequent request.

pushPath—"SipServletRequest.pushPath" method includes a "Path" header in a SIP REGISTER request when acting in the role of a User Agent Client or a Proxy. The SIP "Path" header is defined in [6]. The method has a single parameter of type "Address" interface that specifies the SIP URI to be used when routing subsequent requests based on the SIP registration.

pushRoute—"SipServletRequest.pushRoute" method includes a SIP "Route" header in the outgoing SIP request when acting as a User Agent Client or proxy server. This then determines the next hop that the SIP request should visit. There are two variations of this method:

- The first variation takes a single parameter that is an of the "SipURI" interface, specifying the SIP URI that should be pushed into the SIP request as a "Route" header.

- The second variation takes a single parameter that is an instance of the "Address" interface, specifying the SIP URI and associated "Address" parameters that should be pushed into the SIP request as a "Route" header.

send—"SipServletRequest.send" method, when invoked, causes the SIP protocol request that has been constructed to be sent to the network. This method is used only by User Agent Client applications.

setMaxForwards—"SipServletRequest.setMaxForwards" method is used by applications to configure the SIP "Max-Forwards" header. The SIP "Max-Forwards" header restricts the number of hops that a request will travel before a loop is detected. Each entity traversed subtracts one from the value of this SIP header. The method takes a single integer value indicating the number of hops to be traversed before reporting a looped request.

setRequestURI—"SipServletRequest.SetRequestURI" method allows applications to set the Request URI of the SIP message. The Request URI of a SIP request indicates the true destination of the request. The method takes a single parameter of type "URI."

Example:

```
SipServletRequest cancel = initialInviteRequest.createCancel();
cancel.send();
```

10.4.3 SipServletResponse

The "SipServletResponse" interface represents a SIP response message (as opposed to a SIP request message). The "SipServletResponse" interface extends the "javax.servlet" interface to provide SIP specific methods and also implements the "SipServletMessage" interface to provide common methods across both "SipServletResponse" and "SipServletRequest" interfaces. The following methods are defined for specific manipulations on a "SipServletResponse" interface instance:

createAck—"SipServletResponse.createAck" method returns an instance of the "SipServletRequest" interface representing a SIP ACK request that is based on the 2xx class response that has been received to a SIP INVITE request. An application can then manipulate the "SipServletRequest" interface instance as a request and then use the "SipServletRequest.send" to dispatch to the network.

createPrack—"SipServletResponse.createPrack" method returns an instance of the "SipServletRequest" interface representing a SIP PRACK request that is based on the SIP reliable provisional (1xx class response) received. An application can then manipulate the "SipServletRequest" interface instance as a request and then use the "SipServletRequest.send" to dispatch to the network.

getChallengeRealms—"SipServletResponse.getChallengeRealms" method returns a string "Iterator" (java.util.Iterator<java.lang.string>) containing a list of all the authentication realms that appeared in the "SipServlet Response" instance. The realm is extracted from the appropriate SIP protocol header that is included as part of the challenge—see [1] for more details on SIP authentication.

getProxy—"SipServletResponse.getProxy" method returns an instance of "Proxy" interface associated with the SIP response message. Using the "Proxy" interface signifies that the application is acting as a SIP proxy server. See the description for the "Proxy" interface in this chapter for more details on functions provided when acting as a SIP proxy server.

getProxyBranch—"SipServletResponse.getProxyBranch" method returns an instance of the "ProxyBranch" interface. The "ProxyBranch" interface is detailed elsewhere in this chapter, and the mechanics behind the branch concept are described in Chapter 3.

getReasonPhrase—"SipServletResponse.getReasonPhrase" returns a string representation of the reason phrase that was included in the "SipServlet Response." For example, in the following 487 SIP response, the reason phrase is in bold type.

```
SIP/2.0 487 Request Terminated
Via: SIP/2.0/UDP
sipservlet_example.com;branch=z9hG483JKSJ8ew9;received=192.
0.2.222
To: Stoffe <kristoffer@sipservlet_example.com > tag=890092834
```

```
From: Chris <chris@sipservlet_example.com >;tag=8327489874
Call-ID: fj8493ijf984ulw94@sipservlet_example.com
CSeq: 1 INVITE
```

getStatus—"SipServletResponse.getStatus" method returns an integer value representing the SIP response code of the message. For example, in the following 487 SIP response, the reason phrase is in bold type.

```
SIP/2.0 487 Request Terminated
Via: SIP/2.0/UDP
sipservlet_example.com;branch=z9hG483JKSJ8ew9;received=192.0.2.222
To: Stoffe < kristoffer@sipservlet_example.com > tag=890092834
From: Chris <chris@sipservlet_example.com >;tag=8327489874
Call-ID: fj8493ijf984ulw94@sipservlet_example.com
CSeq: 1 INVITE
```

setStatus—"SipServletResponse.setStatus" method allows an application to set the status code that is present in a SIP response. There are two variations of this method:

- The first variation has a single integer parameter that sets the status code for the SIP response.
- The second variation has two parameters. The first parameter is an integer that sets the status code for the SIP response. The second is a string which sets the reason phrase for the SIP response.

getRequest—"SipServletResponse.getRequest" method returns an instance of the "SipServletRequest" object that was used to generate the original request that resulted in the SIP response being processed.

isBranchResponse—"SipServletResponse.getRequest" returns a Boolean representing whether a SIP response arrived on the recently introduced Proxy Branch mechanism. A value of "true" is returned if the response arrived on a Proxy Branch, while "false" is returned if it did not arrive on a Proxy Branch. More details on the Proxy Branch mechanism are contained in this section and in Chapter 3.

send—"SipServletResponse.send" method is used to dispatch a SIP response message once the application is ready. Using this method signifies that the application is acting in the role of a SIP User Agent Server (UAS).

sendReliably—"SipServletResponse.sendReliably" method allows an application to send a SIP provisional response in the range of 101 to 199 reliably, as specified in RFC 3262 [8]. An application should only really use this method if it is certain that the upstream User Agent Client (UAC) supports this extension. An application can obtain the list of SIP extensions supported by the container from the Servlet Context attribute "javax.servlet.sip .supported" using the "getAttribute" method. Version 1.0 of the SIP Servlet

API provided the "javax.servlet.sip.100rel" Servlet Context attribute to determine support, but this is no longer the favored mechanism.

Example:

```
if (resp.getStatus() == SipServletResponse.SC_OK) {
 if( "INVITE".equals( resp.getMthod() ) ) {
  SipServletRequest ack = resp.createAck();
  ack.send();
 } else {
 // We are done, but lookout for SUBSCRIBE & REFER :-)
}
```

10.4.4 Address

The "Address" interface is used to conveniently represent a common construct that is used in multiple SIP headers (e.g., the SIP "To," "From," and "Contact" headers). The common construct consists of a URI, an optional display name, and a set of associated header parameters (note that is header parameters and not URI parameters that appear within "<" and ">."

For example:

```
Contact: "Stoffe" < sip:123456@sipservlet_example.com;user=phone >;
foo=890092834
```

In this example, SIP "Contact" header, the "user=phone" parameter is a SIP URI parameter, while the "foo=890092834" parameter is considered a "Contact" header parameter. The address object is defined in the specification as having the following form:

```
Address = (name-addr | addr-spec) *(SEMI generic-param)
```

So, the object includes the SIP URI part and display name as defined by RFC 3261 [1], using the "name-addr" and "addr-spec" constructs as well as the associated header parameters, as defined by the "generic-param" inclusion.

The "Address" interface defines the following methods:

clone—"Address.clone" method returns an object representation of the "Address" interface instance containing all the components previously mentioned (display name, URI, and header) parameters. The only exception is the "Tag" parameter, defined in RFC 3261 [1], which is not cloned, because it violates the SIP protocol to reuse.

equals—"Address.equals" method returns a Boolean value indicating whether two "Address" interface instances are equal. A value of "true" indicates that the two "Address" interface instances are equal, while "false" means they are not. The URI and its parameters are compared as specified in RFC 3261 [1]. The header parameters are also compared the same way that URI parameter comparison is, as specified in RFC 3261 [1]. The optional display name is not used in the comparison. The method takes a single parameter of type "java'lang'Object," which represents the "Address" instance that is to be compared.

getDisplayName—"Address.getDisplayName" returns an object of type string representing the display name extracted from the "Address" interface instance. Here, for example, is an instance of an "Address" interface that would return the value "Stoffe" on calling this method:

```
"Stoffe" < sip:123456@sipservlet_example.com;user=phone >;
foo=890092834.
```

setDisplayName—"Address.setDisplayName" method allows the application to set the display name part of an "Address" interface instance. The method has a single string parameter that is used to specify the display name.

getExpires—"Address.getExpires" method returns the value of the SIP "Expires" header parameter when present. A value of "–1" is returned if the header is not present in the "Address" interface instance.

setExpires—"Address.setExpires" method allows an application to explicitly set the "Expires" header parameter on an "Address" interface instance. The method has a single parameter of type integer that is used to represent the number of seconds that are to be set in the SIP "Expires" header.

getQ—"Address.getQ" method returns an object of type float representing the value of the "q" header parameter defined in RFC 3261 [1]. The "q" value parameter is used to set preference when multiple locations are present in SIP REGISTER requests.

setQ—"Address.setQ" method allows an application to explicitly set the value of the "q" parameter specified in RFC 3261 [1]. The method has a single value of type float, which is then used to populate the "q" header parameter in the "Address" instance.

isWildCard—"Address.isWildCard" method allows an application to check for a special "*" wild card that may appear in a SIP "Contact header" as defined in RFC 3261 [1]. The use of the "*" wild card in a SIP REGISTER request acts as a "remove all" current bindings operation. This method allows an application to check for this special case in an "Address" interface instance. The method returns a Boolean value, with a value of "true" indicating that this is the special "*" case, while "false" indicates that it is not.

getURI—"Address.getURI" method provides the application with only the URI part of an "Address" instance. The method returns an instance of the "URI" interface introduced in this chapter.

setURI—"Address.setURI" method allows an application to set the URI part of an "Address" instance. The method has a single parameter that is an instance of the "URI" interface containing the chosen value.

toString—"Address.toString" method provides an application with the ability to convert an instance of the "Address" interface into a string object.

Example:

```
Address addr = sipFactory.createAddress("\"Kristoffer
Gronowski\"<sip:stoffe@sipservlet.net>");
addr.setExpires(3600);
req.addAddressHeader("m",addr);
```

10.4.5 Parameterable

The "Parameterable" interface is a newly introduced convenience method that allows easy manipulation of a SIP header field with optional parameters. This SIP Servlet specification defines a parameterable as follows:

```
field-name: field-value * (;parameter-name[=parameter-value])
```

An example of a parameterable would be the following SIP "Via" header:

```
Via: SIP/2.0/UDP
sipservlet_example.com;branch=z9hG483JKSJ8ew9;received=192.0.2.222
```

In this example, the "field-name" is equal to "Via." The "field-value" is equal to "SIP/2.0/UDP sipservlet_example.com." The "parameter-name" and "parameter-value" of the parameterables equal "branch=z9hG483JKSJ8ew9" and "received=192.0.2.222."

The "Parameterable" interface has the following methods:

clone—"Parameterable.clone" method returns a "java.lang.Object" instance containing a copy of the "Parameterable" interface instance.

equals—"Parameterable.equals" method compares two instances of the "Parameterable" interface to identity if they are the same. The method returns a Boolean value. The value of "true" indicates that the instances of the "Parameterable" interface are the same, while "false" indicates they are not the same.

getParameter—"Parameterable.getParameter" method allows an application to specify and retrieve the value of a parameter that is present in a "Parameterable" interface instance. The method has a single string parameter that specifies the name of the parameter to be returned. The method returns the parameter value as a string.

getParameterNames—"Parameterable.getParameterNames" method returns a "java.util.Iterator<java.lang.String>" object containing a list of all parameter names that are present in a "Parameterable" instance.

getParameters—"Parameterable.getParameters" method returns a "java.util.Set <java.util.Map.Entry<java.lang.String, java.lang.String>" object containing a view of the name-value parameter mappings contained in the "Parametable" interface instance.

getValue—"Parameterable.getValue" method returns the field value of a "Parameterable" interface instance as a string object.

setValue—"Parameterable.setValue" method allows an application to set the field value of a "Parameterable" interface instance. The method has a single parameter of type string that represents the value that the field should be set.

removeParameter—"Parameterable.removeParameter" method allows an application to remove a specified parameter from a "Parameterable" interface instance. The method has a single string parameter that specifies the parameter that is to be removed.

setParameter—"Parameterable.setParameter" method allows an application to set the value of a parameter contained in an instance of the "Parameterable" interface. The method has two parameters. The first parameter is a string representing the name of the parameter to be set. The second parameter is also a string parameter representing the value of the parameter being set. If the parameter already exists with the "Parameterable" instance, then it is replaced; otherwise, it is added.

Example:

```
Parameterable p = sipFactory.createParameterable("*;audio");
p.setParameter("explicit","");
p.setParameter("q","1.0");
req.setParameterableHeader("Accept-Contact",p);
```

10.4.6 SipURI

The "SipURI" interface is used to represent both "sip:" and "sips": URIs used in the SIP protocol. The SIP URI scheme is used extensively in the protocol in a number of places for various routing and signaling purposes. It is for this reason that this convenience interface allows for ease of manipulation. The construct can be closely compared to e-mail in that an identifier consists of a user part and a host part, for example, chris@example.com. The user part is equal to "chris," and the host part is equal to "example.com." A SIP URI additionally has parameters that are separated using ";" followed by headers that are separated by "?". The SipURI interface consists of the following methods:

equals—"SipURI.equals" method provides the ability to compare two instances of the "SipURI" interface. It has a single parameter of type "java.lang.object" that represents the SIP URI to be compared. The method returns a Boolean value of "true" if the two SIP URIs are equal and "false" if they are not. The rules for comparing two SIP URIs are described in RFC 3261 [1].

toString—"SipURI.toString" method converts an instance of the "SipURI" interface into a string.

getHeader—"SipURI.getHeader" method provides the value of a specified header. A SIP URI can contain header parameters that are included using a "?". For example, "sip:chris@example.com?Subject=Conference" illustrates a SIP URI with a "Subject" header equal to "Conference." If in the future a SIP request is constructed based on this URI, the header will also be included. The method has a single string parameter specifying the name of the header. A string is then returned, conveying the value of the header. From our previous example, "header_value.getHeader("Subject")" would return the value "Conference."

setHeader—"SipURI.setHeader" method allows an application to set the value of a header in a SIP URI. The method has two parameters. The first parameter is a string representing the name of the header to be set in the "SipURI" instance. The second parameter is a string representing the value of the header. For example, "SipURI_instance.setHeader("Subject", "Conference")" would result in a "Subject" header being added to the "SipURI" instance with a value of "Conference."

removeHeader—"SipURI.removeHeader" method enables an application to remove a header from an instance of "SipURI" interface. The method has a single parameter of type string that specifies the header that is to be removed. For example, "SipURI_instance.removeHeader("Subject")" would result in the "Subject" header being removed from the "SipURI" instance.

getHeaderNames—"SipURI.getHeaderNames" method returns an Iterator "java.util.Iterator<java.lang.String>" listing all the header names that are present in an instance of the "SipURI" interface.

getHost—"SipURI.getHost" method provides an application with the host part of a "SipURI" instance (i.e., the domain part that appears after the "@" symbol). The method returns a string of the host part.

setHost—"SipURI.setHost" method allows an application to set the host part of a "SipURI" interface instance. The method has a single parameter of type string that specifies the host part of a SIP URI.

getLrParam—"SipURI.getLrParam" method informs an application whether the "lr" parameter is set on a "SipURI" interface instance. The "lr" parameter was introduced in a revision of the SIP protocol to determine a new mechanism called "loose route." For more information on the "loose route" mechanism and the "lr" parameter, take a look at RFC 3261 [1]. This method returns a Boolean indicating whether the "lr" parameter is present in the "SipURI" instance. A value of "true" means the "lr" parameter is present, and "false," that it is not.

setLrParam—"SipURI.setLrParam" method enables an application to set the presence of the "lr" parameter in an instance of the "SipURI" interface. The method has a single parameter of type Boolean. A value of "true" sets the "lr" parameter to be present, while a value of "false" sets the "lr" parameter to be not present.

getMAddrParam—"SipURI.getMAddrParam" method returns the value of the "maddr" parameter. The "maddr" parameter is defined in RFC 3261 [1]. The method returns a string representation of the "maddr" parameter.

setMAddrParam—"SipURI.setMAddrParam" method enables an application to set the value of the "maddr" parameter in a "SipURI" interface instance. The method has a single parameter of type string that specifies the value to be used.

getMethodParam—"SipURI.getMethodParam" method returns the value of the "method" parameter. The "method" parameter is defined in RFC 3261 [1]. This method returns a string containing the value of the "method" parameter from the "SipURI" instance.

setMethodParam—"SipURI.setMethodParam" method enables an application to set the value of the "method" parameter in a "SipURI" instance. The method has a single string parameter that specifies the value to be used for the "method" parameter.

getPort—"SipURI.getPort" method provides the network port associated with the "SipURI" instance. A "SipURI" instance has an optional port configuration, which can be seen in more detail in RFC 3261 [1]. The method returns an integer representing the port value for the "SipURI" interface instance.

setPort—"SipURI.setPort" method allows an application to configure the port of a "SipURI" instance. The method has a single parameter of type integer that specifies the value of the port.

getTransportParam—"SipURI.getTransportParam" method provides the contents of the "transport" parameter, which is defined as part of a SIP URI in RFC 3261 [1]. The method returns a string representation of the "transport" parameter.

setTransportParam—"SipURI.setTransportParam" method enables an application to set the value of the "transport" parameter for a "SipURI" interface instance. The method has a single string parameter specifying the value for the "transport" parameter.

getTTLParam—"SipURI.getTTLParam" method provides the contents of the "ttl" parameter, which is defined as part of a SIP URI in RFC 3261 [1]. The method returns an integer representation of the "ttl" parameter.

setTTLParam—"SipURI.setTTLParam" method enables an application to set the value of the "ttl" parameter for a "SipURI" interface instance. The method has a single integer parameter specifying the value for the "ttl" parameter.

getUser—SipURI.getUser" method extracts the user part of a SIP URI (before the "@" symbol) as defined in RFC 3261 [1]. The method returns a string representation of the user part of a "SipURI" instance.

setUser—"SipURI.setUser" method enables an application to set the user part of a SIP URI (before the "@" symbol) as defined in RFC 3261 [1]. The method has a single string parameter specifying the value of the user part of a "SipURI" interface instance.

getUserPassword—"SipURI.getUserPassword" method provides the password of the SIP URI if it is set (null if it is not set) The method returns a string representation of the password part of a "SipURI" instance.

setUserPassword—"SipURI.setUserPassword" method enables an application to set the value of the password part for a SIP URI. The method has a single parameter of type string specifying the password part of a "SipURI" instance.

getUserParam—"SipURI.getUserParam" method provides the contents of the "user" parameter that is defined as part of a SIP URI in RFC 3261 [1]. The method returns a string representation of the "user" parameter.

setUserParam—"SipURI.setUserParam" method enables an application to set the value of the "user" parameter for a "SipURI" interface instance. The method has a single string parameter specifying the value for the "user" parameter.

isSecure—"SipURI.isSecure" method informs the application whether the instance of the "SipURI" interface is secure (by definition, if it is of type "sips," it is secure, and if of type "sip," it is not). The method returns a Boolean, with a value of "true" indicating it is secure and "false indicating" it is not.

setSecure—"SipURI.setSecure" method enables an application to set whether an instance of the "SipURI" interface is secure (by definition, this sets the scheme to either "sip" or "sips"). The method has a single Boolean parameter whereby the value of "true" specifies that it should be secure (sips), and "false," that it should not (sip).

Example:

```
SipURI uri = sipFactory.createSipURI( "stoffe", "sipservlet.net" );
uri.setSecure(true);
uri.setPort(35061);
//Would produce "sips:stoffe@sipservlet.net:35061'
```

10.4.7 TelURL

This book has already looked at the "SipURI" interface for representing the SIP protocol. For legacy reasons, it is also desired within the SIP protocol to be able to represent traditional PSTN telephone numbers. While it can be envisioned that, in the future, a person would be located using an e-mail-style SIP URI format, the use of traditional phone numbers will be around for a long time to come, and so they need representation in the SIP protocol and SIP Servlet API. For this reason, the "tel:" URI scheme was defined in RFC 3966 [2]. For more information on valid use of the "tel:" URI scheme in the SIP protocol, see RFC 3261 [1]. The following methods are defined for the "TelURL" interface.

equals—"TelURL.equals" method enables an application to compare an instance of the "TelURL" interface with another instance. The method has a single parameter of type "java.lang.Object," which represents the other Tel URL that is being compared. The method returns a Boolean value indicating whether the two instances of the "TelURL" interface are equal.

getPhoneContext—"TelURL.getPhoneContext" method returns the value of the "Phone-Context" parameter for the instance of the "TelURL" interface or returns null if it's a global number. The method returns a string representation of the value from the "Phone-Context" parameter.

getPhoneNumber—"TelURL.getPhoneNumber" method returns the value of the phone number associated with a "TelURL" interface instance. The method returns a string value representing the telephone number.

setPhoneNumber—"TelURL.setPhoneNumber" method enables an application to set the telephone number for an instance of the "TelURL" interface. The method has two variations:

- The first instance has a single parameter. The parameter is of type string and represents the phone number to be set for the "TelURL" interface instance: for example, "TelURL_instance.setPhoneNumber ("123456")."

- The second instance has two parameters. The first parameter is of type string and represents the phone number to be set for the "TelURL" interface instance. The second parameter is of type string and represents the phone-context of the telephone number: for example, "TelURL_instance.setPhoneNumber("123456," "example.com")."

isGlobal—"TelURL.isGlobal" method informs an application whether a telephone number is global. It returns a Boolean, with a value of "true" indicating that the telephone number is global and a value of "false" that it is not global.

toString—"TelURL.toString" method returns a string representation of the "TelURL" instance.

Example:

```
TelURL tel = (TelURL) sipFactory.createURI("tel:+1-555-1234-567");
if( tel.isGlobal() ) {
 log("We should end up here since the number started with +1");
}
```

10.4.8   URI

The "URI" interface provides a base for all URI manipulations that span both the "sip:," "sips," and "tel" schemes. The generic definition for a URI is provided in RFC 2396 [9]. The "URI" interface has the following method:

clone—"URI.clone" method enables an application to produce an exact replica of a "URI" instance. The method returns a new instance of the "URI" interface.

equals—"URI.equals" method compares two instances of the "URI" interface. It has a single parameter of type "java.lang.Object," which represents the second "URI" instance to be compared. The method returns a Boolean indicating whether the two "URI" interface instances are the same.

getParameter—"URI.getParameter" method returns the value of a specified URI parameter. The method has a single string parameter specifying the parameter that is to be retrieved. The method returns a string containing the value of the specified parameter. For example, using the URI "chris@ example.com;uri-param=1234" as an example and calling the URI_instance .getParameter("uri-param") would return the value "1234."

setParameter—"URI.setParameter" method enables an application to configure a URI parameter. The method has two parameters. The first parameter is of type string and represents the name of the parameter to be set. The second parameter is also a string specifying the value of the parameter.

removeParameter—"URI.removeParameter" method enables an application to remove a parameter from a "URI" interface instance. The method has a single string parameter that indicates the parameter to be removed.

getParameterNames—"URI.getParameterNames" method returns a "java.util .Iterator<java.lang.String>" listing the names of all the parameters for the "URI" interface instance.

getScheme—"URI.getScheme" method provides the scheme for the "URI" interface instance. The method returns a string with a value of "sip," "sips," or "tel."

isSipURI—"URI.isSipURI" method enables an application to determine whether an instance of the "URI" interface is SIP based. The method returns a Boolean whereby a value of "true" means that it is SIP based (the scheme was either "sip:" or "sips:") and a value of "false" means that is was some other scheme.

toString—"URI.toString" method enables an application to convert a instance of the "URI" interface into a string representation.

Example:

```
SipServletRequest req;
URI aUri = req.getRequestURI() ;
if( aUri.isSipURI() ) {
 SipURI aSipUri = (SipURI) aUri ; //a safer cast
}
```

10.5 Timer Service

A SIP Servlet container also provides a basic timer service that applications can use and receive notification events when a specific timer expires. The timer service provided by the container allows for data to be stored with a timer that can then be retrieved at a later period. The timer service also allows for varying levels of granularity of timer configuration to incorporate multiple occurrences. The overall timer service consists of three interfaces, which will be discussed in the remainder of this chapter.

10.5.1 TimerService

The "TimerService" interface is the primary entry point for applications wishing to use the SIP Servlet timer functionality. This interface allows for the setting of timers and will at some later time be notified of expiration (using the "Timer Listener" interface described later in this section). The "TimerService" interface should be offered by containers using the "javax.servlet.sip.TimerService" Servlet Context attribute. The "TimerService" interface has the following method:

createTimer—"TimerService.createTimer" method is the mechanism used to configure a timer in a SIP Servlet container. The method returns an instance of the "ServletTimer" interface that represents a timer that has been installed in the container. The "ServletTimer" interface is described in more detail later in the section. The method has two variations:

- The first variation has four parameters that are configured when installing new timer:
  - ○ The first parameter is the instance of the "SipApplicationSession" interface that is to be associated with the timer.
  - ○ The second parameter is of type long and represents the delay in milliseconds before the timer expires.
  - ○ The third parameter is of type Boolean and signifies whether the timer is to be persistent during a SIP Servlet container failure or to shut down. If set to "true," then the timer is persistent and is reinstalled on a restart. If set to "false," the timer is not reinstalled on a restart.
  - ○ The fourth parameter is of type "java.io.Serializable" and represents any information that the application would like to be delivered when the timer fires and a notification is sent to the "Timer Listener" interface (which is discussed later in this chapter).
- The second instance has six parameters that are configured when installing a new timer:
  - ○ The first parameter is the instance of the "SipApplicationSession" interface that is to be associated with the timer.
  - ○ The second parameter is of type long and represents the delay in milliseconds before the timer expires.
  - ○ The third parameter is of type long and represents the time milliseconds between successive timer expirations.

○ The fourth parameter is a Boolean that when "true" specifies that the timer is scheduled in a fixed-delay mode. When "false," the timer is scheduled in fixed-rate mode. Fixed-rate mode is defined in the SIP Servlet specification as follows: "In fixed-rate execution, each execution is scheduled relative to the scheduled execution time of the initial execution. If an execution is delayed for any reason (such as garbage collection or other background activity), two or more executions will occur in rapid succession to 'catch up.' In the long run, the frequency of execution will be exactly the reciprocal of the specified period (assuming the system clock underlying Object.wait(long) is accurate)" [1]. Fixed-delay mode is defined in the SIP Servlet specification as follows: "In fixed-delay execution, each execution is scheduled relative to the actual execution time of the previous execution. If an execution is delayed for any reason (such as garbage collection or other background activity), subsequent executions will be delayed as well. In the long run, the frequency of execution will generally be slightly lower than the reciprocal of the specified period (assuming the system clock underlying Object.wait(long) is accurate)" [1].

○ The fifth parameter is of type Boolean and signifies whether the timer is to be persistent during a SIP Servlet container failure or shut down. If set to "true," then the timer is persistent and is reinstalled on a restart. If set to "false," the timer is not reinstalled on a restart.

○ The sixth parameter is of type "java.io.Serializable" and represents any information that the application would like to be delivered when the timer fires and a notification is sent to the "Timer Listener" interface (which is discussed later in this chapter).

10.5.2 ServletTimer

The "ServletTimer" interface is the result of a timer being created using the "TimerService" interface and contains specific timer information previously created. The "ServletTimer" interface has the following methods:

cancel—"ServletTimer.cancel" method cancels the timer regardless of whether it's a repeating or one-occurrence timer.

getApplicationSession—"ServletTimer.getApplicationSession" method returns the instance of the "SipApplicationSession" associated with the timer.

getId—"ServletTimer.getId" method returns a string containing the identifier assigned to the specific timer task.

getInfo—"ServletTimer.getInfo" method enables the application to retrieve application-specific data it has stored in association with the timer on creation. The method returns the information as a "java.io.Serializable" object.

getTimeRemaining—"ServletTimer.getTimerRemaining" returns the amount of time in milliseconds remaining before the timer fires. The method returns the timer as a long.

scheduleExecutionTime—"ServletTimer.scheduleExecutionTime" provides the most recent execution of the timer. It is typically used with the previously described fixed-delay mode, which allows execution times to drift over time. The result is returned a long.

Example:

```
public class TimingServlet extends javax.servlet.sip.SipServlet {

  @Resource
  TimerService ts;

  ServletTimer st;
  st = ts.createTimer(sas, 3600, false, "Something for the
TimerListener");
  /// We changed our mind and don't need the timer anymore
  st.cancel();
```

10.5.3   TimerListener

The "TimerListener" is a listener interface that is implemented by applications wanting to be notified when a timer fires. The "TimerListener" interface has the following method:

timeout—"TimerListener.timeout" method is called by the container when a timer expires to notify the application. The method provides the application with a single parameter of type "ServletTimer," which contains all the appropriate information relating to the timer that has fired.

Example:

```
@SipListener

public class TimedSipServlet extends SipServlet implements
TimerListener{
```

```
public void timeout(ServletTimer st) {
 //Write code that should happen when timer is fired.
 }

}
```

References

[1] Rosenberg, J. et al., "SIP: Session Initiation Protocol," RFC 3261, Internet Engineering Task Force, June 2002.

[2] Schulzrinne, H., "The Tel URI for Telephone Numbers," RFC 3966, Internet Engineering Task Force, December 2004.

[3] Mahy, R., and D. Petrie, "The Session Initiation Protocol (SIP) 'Join' Header," RFC 3911, Internet Engineering Task Force, October 2004.

[4] Mahy, R., B. Biggs, and R. Dean, "The Session Initiation Protocol (SIP) 'Replaces' Header," RFC 3891, Internet Engineering Task Force, September 2004.

[5] Schulzrinne, H., D. Oran, and G. Camarillo, "The Reason Header Field for the Session Initiation Protocol (SIP)," RFC 3326, Internet Engineering Task Force, December 2002.

[6] Willis, D., and B. Hoeneisen, "Session Initiation Protocol (SIP) Extension Header Field for Registering Non-Adjacent Contacts," RFC 3327, Internet Engineering Task Force, December 2002.

[7] SIP Servlet Specification, Version 1.1, JSR 289, Java Community Process, August 2008.

[8] Rosenberg, J., and H. Schulzrinne, "Reliability of Provisional Responses," RFC 3262, Internet Engineering Task Force, June 2002.

[9] Berners-Lee, T., R. Fielding, and L. Masinter, "Uniform Resource Identifiers (URI): Generic Syntax," RFC 2396, Internet Engineering Task Force, August 1998.

About the Authors

Chris Boulton is the chief technology officer (CTO) for NS-Technologies and was formerly a technical research specialist working on next-generation solutions for Avaya. He has been an active participant in the Internet Engineering Task Force (IETF) for more than six years. Dr. Boulton has authored and participated in numerous specifications related to the Session Initiation Protocol (SIP) and its related extensions. This includes work on Network Address Translator (NAT) traversal of SIP, Media Server Control using SIP, and VoIP conferencing. He represented Avaya (and Ubiquity Software Corporation before it was purchased by Avaya) on the SIP Servlet 1.1 expert group and has worked closely with the technology since its conception (SIP Servlet 1.0).

Dr. Boulton is also a member of the JSR 309 Media Server API expert group in the Java Community Process (JCP), is on the Technical Board of Advisors for the VoIP Security Alliance (VOIPSA), and has published numerous white papers related to VoIP and related technologies.

Kristoffer Gronowski is a senior software architect at Ericsson Research focusing on empowering development and evolution of communication services. He has worked most of his career in product development, and most recently he has been driving the architecture of the open source SailFin SIP container project from Ericsson. The initial donation of code made by Ericsson originated from the platform product Ericsson Application Server, of which he was one of the three founders and the architect. At the same time, Dr. Gronowski has been contributing to the JSR 289 standard as member of the JCP expert group, where he met Dr. Boulton, and thus the requirement for this particular book was formed between them. Before joining Ericsson, Dr. Gronowki was working for a startup and was a pioneer in the SIP field at Hotsip. There the foundation of multimedia services such as presence, VoIP, and messaging were not only standardized but also proven in solid implementation; this was where Dr. Gronowski gained most of his experience and identified the need to simplify the environment in order to make the technology developer-friendly.

Index

Recent Titles in the Artech House Telecommunications Series

Vinton G. Cerf, Senior Series Editor

*Videoconferencing and Videotelephony: Technology and Standards,
Second Edition,* Richard Schaphorst

Visual Telephony, Edward A. Daly and Kathleen J. Hansell

Wide-Area Data Network Performance Engineering, Robert G. Cole and
Ravi Ramaswamy

Winning Telco Customers Using Marketing Databases, Rob Mattison

WLANs and WPANs towards 4G Wireless, Ramjee Prasad and
Luis Muñoz

World-Class Telecommunications Service Development, Ellen P. Ward

For further information on these and other Artech House titles,
including previously considered out-of-print books now available through our
In-Print-Forever® (IPF®) program, contact:

Artech House
685 Canton Street
Norwood, MA 02062
Phone: 781-769-9750
Fax: 781-769-6334
e-mail: artech@artechhouse.com

Artech House
46 Gillingham Street
London SW1V 1AH UK
Phone: +44 (0)20 7596-8750
Fax: +44 (0)20 7630-0166
e-mail: artech-uk@artechhouse.com

Find us on the World Wide Web at: www.artechhouse.com